Alpaca Keeping

Raising Alpacas – Step by Step Guide Book… farming, care, diet, health and breeding

By Harry Fields

Foreword

When I was researching the history of alpacas to write this book, I discovered that in Andean mythology, the animals are linked to the Earth Mother, "Pachmana." The belief is that alpacas are on loan to us from the gods and are only here for us so long as we respect them and care for them properly.

The alpaca was so important to the native peoples of South America that when the Spanish Conquistadores arrived, they found an entire civilization predicated on the importance of textiles. Even the wealth of prosperous individuals was recorded in a tied pattern of knots in fiber.

Alpaca fleece was the most prized of these fibers, literally being used as currency. The Spanish saw only the gold, silver, and precious stones that were also used liberally by the native peoples and it was for those commodities that the great civilizations were plundered and decimated.

In the process, however, the Spanish also destroyed the truth wealth of the South American world, the native alpacas and llamas that were slaughtered and left to rot in the countryside. Only a few of the animals were secreted away and hidden in remote regions, kept safe there until the fiber was "re-discovered" by European textile manufacturers during the Industrial Revolution of the 19th century.

It has been said that in their native region, alpacas live closer to heaven than any other animal on earth. They are a respectful species, doing no damage to the earth with their padded feet and eating only the tops of the grass on which they graze, allowing the pasture to renew as they move on.

Alpacas are gentle herd animals, amiable in their relationship with their human keepers, and placid in their adaptability to different climates and landscapes. For this reason, they are now raised around the world, cultivated in a myriad of ownership models from full-scale breeding operations to life as family pets. There is, literally, no one "right" way to own an alpaca.

In fact, my own experience with the animals began as an investment. Because the American market is still breeder based, meaning the foundation herd nationwide is still being built toward a transition to a full fiber market (which could require as many as a million animals) herdsires and dams are very expensive.

A few years ago, a group of friends and I invested in a good quality herdsire, seeing a return on our investment in stud fees. One member of our group lived in the country, and it was there that "Bentley" was kept in between his working "appointments."

The first time I met the handsome animal, which looked for all the world like a cross between a long-necked teddy bear and a poodle, I was captivated by his calm gaze and willingness to interact. Although some alpacas are not receptive to petting, Bentley was not among their number and was always willing to have his chin or ears scratched.

Foreword

From the perspective of an investor, I began to learn more about alpacas, and to this day wish that I could live in the country and be more actively involved in the daily care and breeding of these animals.

My purpose in pulling together this book is to provide a solid overview of the alpaca industry and the husbandry these animals require to help people curious about entering this growing field. As you will see, alpaca ownership is so flexible that even individual textile artisans can keep a pair of alpacas, enjoying both their company and their superb fiber.

I find it almost impossible not to smile when I see an alpaca or better yet, have the chance to interact with one. Even if you come to the end of this book and decide that the world of alpacas is not for you, I think you will have a new appreciation for them.

Alpacas are indeed so unique in their versatility, usefulness, and disposition, it's not hard to believe that they just might be on loan to us from some divine source — one to which we should be grateful for the gracious gift.

Table of Contents

Table of Contents

Table of Contents

Table of Contents

Chapter 1 - Introduction to Alpacas

The alpaca (*Vicugna pacos*) is camelid indigenous to South America that has been domesticated for the cultivation of its fiber. Camelids are members of the *Camelidae* family, which includes llamas, vicunas, guanacos, dromedaries, and Bactrian camels.

Camelids

In physical form, camelids are large mammals. Many are big enough to serve as beasts of burden, but the alpaca is not among that group. These animals are herbivores and quasi ruminants with three-chambered stomachs.

Four of these creatures look very much alike: alpacas, llamas, vicunas, and guanacos.

- **llama** – Although very similar to the alpaca, llamas are larger, with longer heads and curving ears. Like the alpaca, however, llamas are only known in their domestic state. Be temperament, they are more aggressive, and are often used as guard animals.

- **vicuna** – Vicunas have long coats that are wooly in texture. They are brown on the back and white on the throat and chest. They live exclusively in South America and have not been domesticated. At the shoulder they stand about 3 feet (75-85) cm tall.

- **guanaco** – Also a wild animal, the guanaco is similar in size to the vicuna, but just slightly larger. They have gray faces, with a body color that ranges from dark cinnamon to light brown. The underside of the body is white. With less than 600,000 guanacos left in the mountainous ranges of South America, they are considered to be vulnerable from a conservation status.

Dromedaries are Arabian and Indian camels, with the traditional "camel" physical appearance and a single hump. The Bactrian camel, which is indigenous to Central Asia, is a shaggy, short, domesticated pack animal with two humps.

Quasi-Ruminants

Their three-phase digestion distinguishes camelids from true ruminants. When camelids eat, the plant matter is first passed to the rumen, where a process of microbial fermentation occurs.

From the rumen, the partially digested food moves to the reticulum, which is filled with a fluid that serves to separate solid and liquid materials from the ingested feed matter.

The solids are clumped together to form a bolus, commonly referred to as a "cud," which is then regurgitated and chewed again before being passed to the real stomach, the abomasum.

Even-Toed Ungulates

Camelids are also even-toed ungulates. Many large mammals are ungulates. They use the tips of their toes or hooves to sustain the bulk of their body weight when they are in motion. This diverse group includes:

- pigs
- peccaries
- hippopotamuses
- camels
- llamas
- alpacas
- deer
- giraffes
- pronghorn antelope
- sheep
- coats
- cattle

Camelids are relatively long-legged animals, standing on two central toes with nails. Some species have surviving third and fourth toes as dew claws.

Alpaca Characteristics

Alpacas are unique in the fiber industry in that their hair naturally grows in 52 natural colors (as classed in Peru), with 12 recognized by the Australian alpaca industry, and 16 by farmers in the United States.

Adult alpacas measure 32-39 inches / 81-99 cm at the withers or shoulder, and about 5 feet / 1.5 meters from the ground at the head. Adults weigh 106-185 lbs. / 48-84 kg.

Unique Behaviors

Alpacas are social animals living in family groups dominated by an alpha male. The lifespan is approximately 20 years. They have a number of interesting and unique behaviors.

Communal Dung Piles

A group of alpacas will choose a spot away from their grazing grounds to create a communal dung pile. Female alpacas tend to all go to the dung pile at once, while males maintain smaller, more individual piles.

This behavior has an added health benefit in that it limits the spread of internal parasites among group members. It also makes pasture maintenance much easer!

Tolerance of Handling

Alpacas have a reputation for being friendly with their human keepers if they receive early socialization and are worked with on a regular basis.

They do, however, have an aversion to being grabbed suddenly, which is in keeping with their role in nature as prey animals. They also don't like to be touched on their feet, lower legs, and abdomens.

Spitting

Camels have a reputation for "spitting," which is also a potential behavior among alpacas. My father nursed a life-long hatred of camels after having been spat upon by one in North Africa during World War II.

With alpacas, the "spit" is actually an acidic blob of partially digested grass brought up from the stomach and hurled at other alpacas. For instance, females refusing a male during mating will spit to signify their displeasure.

The good news is that alpacas rarely if ever spit at humans!

Vocalizations

Alpaca groups communicate with a fairly complex vocabulary of sounds, emitting high-pitched shrieks to indicate danger and a "wark" when excited.

Clicking or clucking indicates a friendly or even submissive reaction, while humming is a comforting and contented sound. Fighting males scream at one another with a cry that is oddly bird like.

Indigenous Range

Alpacas are native to the high Andes Mountains of the southern part of Peru, northern Bolivia, northern Chile, and Ecuador.

They live year round at altitudes of 11,500-16,000 feet / 3,500-5,000 meters above sea level. There are two species, the Suri and the Huacaya.

This region, called the Altiplano, Andean Plateau, or Bolivian Plateau is a semi-arid to arid region with a cool and humid climate.

The annual temperature varies from 37-53 F / 3-12 C and the average annual rainfall is 7.8-31.5 inches / 200-800 mm.

Some parts of the area can see seasonal lows of -4F / -20 C, and snowfall typically occurs in the north between April and September.

Interestingly, however, alpacas are amazingly adaptable to different climates with appropriate help from their keepers. Depending on location, your animals will require shelter for warmth and shade and potentially cooling measures like fans and cool spray misters, but they will thrive under a wide variety of conditions.

This adaptability has allowed the alpaca to be carried well out of its native range in South America, and the animals are now raised successfully around in the world.

There are thriving alpaca farms throughout North America, in the UK, Europe, Australia, and New Zealand among other locales.

Huacaya and Suri

There are two types of alpacas Huacaya and Suri. Approximately 95% of the alpacas you will encounter are Huacayas.

They have crimped fiber that grows perpendicular to their skin creating the appearance of a fluffy teddy bear. Even in its raw condition, the fiber is pleasing to the touch, soft, and very warm.

Suri Alpacas look like the reggae singers of the animal world. Their long, shiny hair hangs in "dreadlocks" that are extremely soft and slightly curly.

The lustre of the Suri fiber is so high it seems to glisten in the sun and feels like fine silk to the touch.

Although both types of alpaca fleece are considered to be luxury textile fibers, of the two, the Suri are the rarest alpacas cultivated. Their fiber commands much higher prices on the international market.

History of Alpacas

The domestication of both alpacas and llamas dates from the period 4000-5000 BC in South America, but it was the Inca who were particularly focused on the production of fine textiles as a symbol of wealth.

During the height of their culture's dominance from 1438 to 1532, they kept meticulous record of flock sizes and engaged in an active breeding program to ensure pure colors in the animals.

Prior to the arrival of the Spanish conquistadores at the border of the Incan empire in 1528, the quality of the fiber produced was far superior to that seen today. Mummified animals were discovered in 1991 that affirm this fact.

The Spanish conquest, however, completely wiped out the results of centuries of selective breeding with the destruction of more than 90% of the existing alpacas and llamas in the region.

In their place, the Spanish introduced their own livestock, forcing the surviving alpacas and llamas into habitats only marginally suited to their survival.

Alpaca fiber was not rediscovered until the 1860s by an English entrepreneur, Sir Titus Salt. The political instability of Central and South America complicated the preservation and cultivation of the remaining animals, however. Radical land reforms in the region brought extant numbers of alpacas down to 2.5 million by 1992.

In the 1980s, large numbers of alpacas began to be exported to the United States, Australia, and Europe. Breeders in these countries and in other parts of the developed world have worked to establish registries with accurate DNA information for the express purpose of improving bloodlines.

On a global basis, the alpaca population is now believed to be around 3 million. The majority of those animals are still found in Peru, Chile, and Bolivia, but alpaca farms can also be found throughout North America, Canada, Australia, New Zealand, and much of Europe.

The presence of these farms does not necessarily mean that a viable alpaca-based fiber industry is present in those countries, however.

Types of Alpaca Markets

The current state of the alpaca industry in any country where the animals are present can either be termed a breeders' market or a fiber market.

The former is a developmental phase focused on accumulating the necessary genetics to make the fiber produced competitive in the world market and a sufficient number of animals to ensure a volume of fiber.

In order for a market to transition from a breeders' climate to a full fiber stage, the national herd size must be large enough to support the activity of the mills. Additionally, there should be fewer people entering the business in relation to the number of animals present in the country.

At that point, alpacas are bought and sold at a price point that actually supports their production. In a breeders' maker, the prices commanded are astronomical, as befits elite stock intended primarily to create a superior genetic foundation.

The United States, for instance, is currently a breeders' market. It is estimated that there would have to be 1 million alpacas in the U.S. before a fiber market could be established. Currently in the Unites States there are 53,000 alpacas.

Alpacas as a Business

There are so many "models" for alpaca ownership that it's impossible to suggest there is any one "right" way to have these creatures in your life. For people who own a few acres and want engaging pets, alpacas are an excellent choice.

Some fiber artists want to own a few animals in order to harvest the fleece for their own projects. More ambitious owners hope to cultivate premier breeding stock commanding high sales and stud fees.

Each of these categories of alpaca ownership comes with unique risks and benefits.

In the United States, for instance, alpacas held as breeding stock for more than a year qualify as a long-term capital gain when sold and are thus considered valuable assets that are 100% insurable.

For a period of five years, the animals are depreciable, giving their owners tax savings while building their herd. Additionally, all expenses related to the alpacas' care are deductible:

- feed
- veterinarian care
- husbandry supplies
- farm equipment
- computers used in flock management
- travel costs for alpaca shows
- show fees
- advertising costs

Depending on the state/regional tax code, alpaca owners may also see a reduction in their real estate taxes. Obviously all tax provisions vary by country, but these kinds of savings can be a major benefit of ownership.

Prices for Alpaca Fleece

Alpaca fiber is a high demand / limited supply textile. As such, owners have little trouble selling fleece to hand spinners or to fiber co-ops or similar sources.

Many purchase spinning equipment and produce their own textile products, which they sell in stores on their alpaca farms. As I said, there is no one "right" way to be in the alpaca business.

As a gauge of fiber prices, a clean, good-quality fleece sells for $3-$5 / £1.8-£3 an ounce (28.34 grams). Therefore, an alpaca fleece weighing 7 lbs. / 3.18 kg would generate around $500 / £298 in income!

Typically that price is more than adequate to cover the annual cost of feed, veterinary care, and maintenance for one alpaca. Unlike most livestock, alpacas can earn their keep quite nicely!

Stud Fees

If you own a fine male "herdsire," you can make an even better income on stud fees, which start in the range of $1,500 / £893 and go up depending on the quality of the animal in question.

Most farms offer three forms of stud service:

- drive-by breeding
- farm breeding
- mobile mating

In *"drive by breeding"* females are transported to the farm where the stud is resident and remain there for a brief period (two-days is standard).

For *"on farm breeding"* the female generally remains at the stud farm for 60-90 days, where she receives routine care and ultrasound to confirm the success of the mating.

With *"mobile mating"* the stud male is transported to the farm where the female is resident. These arrangements are generally limited by distance, and the stay will also be brief.

Obviously given the degree of associated services, each level of stud service is priced differently.

Alpaca Pedigrees

Bloodlines are extremely important in the alpaca industry, with stud males having the greatest influence on programs to improve fiber quality.

The largest alpaca pedigree registry in the world, Alpaca Registry, Inc. (ARI) is located in Lincoln, Nebraska. It is a non-profit entity created by the Alpaca Owners Association, Inc. (AOA) in 1988 for the benefit of U.S. breeders, which has since expanded to include registration of alpacas in Canada and throughout the world.

The ARI database includes the genealogy of approximately 250,000 alpacas and is based on advanced DNA technology to protect the existing gene pool. Every registered animal must be validated as the offspring of two ARI registered parents.

The Certificates of Registration issued display up to five generations, which is a tremendous asset for breeders to verify the integrity of their investment in breeding stock.

Other alpaca registries and breed societies around the world include:

Alpaca Owners Association, Inc. (AOA)
alpacainfo.com

Australian Alpaca Association (AAA) & (IAR) Registry

www.alpaca.asn.au

Australasian Alpaca Breeders Association Inc. (AABA)

www.aaba.com.au

Lama and Alpaka Register (Austria)

www.lamas.at

Canadian Llama and Alpaca Association and Registry (CLAA)

www.claacanada.com

Alpaca Canada

www.alpacainfo.ca

European Suri Association

www.europeanSuriassociation.com

Alpaca Breeders of Finland

www.alpakkakasvattajat.fi

Alpagas et Lamas de France & Registry (l'AFLA)

www.alpagas-lamas-france.org

Alpaka Zucht Verband Deutschland e.V. & Registry (AZVD)

www.alpaka.info

Societa Italiana Alpaca (SIA)

www.sialpaca.it

Alpaca Association of New Zealand (AANZ) & (IAR) Registry

www.alpaca.org.nz

The Norwegian Alpaca Association

www.alpakkaforeningen.no

International Alpaca Association (IAA) (Peru)

www.aia.org.pe

South Africa Alpaca Breeding Society

www.alpacasociety.co.za

Llama and Alpaca Registries Europe (LAREU) (Switzerland)

www.lareu.org

Alpaka Verein Schweiz (VLAS), Alpaca Association Switzerland

www.vlas.ch

The British Alpaca Society and British Alpaca Registry (BAS)

www.bas-uk.com

Suri Network

www.Surinetwork.org

(Please note that the addresses provided above were extant at the time of this writing in mid-2014, but due to the changing nature of the Internet, no guarantee of their validity can be made for the future.)

How Much Do Alpacas Cost?

Putting a price on alpacas is extremely difficult. Perceptions of the value and quality of any animal are largely dependent on the free market and on the buyers and sellers.

For the purpose of examples, the following prices are good approximations for mid-2014.

- A gelded or intact but non-breeding male: $300 - $500 / £180 - £300

- A young female that has already given birth to at least one live offspring: $2,500 - $10,000 / £1495 - £5975

- Older females (3-5 years) with superior conformation and fleece, excellent bloodlines, and a record of performance at shows: $15,000 / £8963.

Herdsire prices will be in excess of $15,000 / £8963 if the individual animal has champion credentials in show and is descended from an excellent bloodline.

In February 2010, a herdsire in the United States sold at auction for $675,000 / £403,346. The record private price for half interest in a herdsire sold in the United States is $750,000 / £448,162.

Estimating Start-Up Farm Costs

Estimating start-up costs for an alpaca farm is quite difficult. It can involve land acquisition and construction as well as the price of the animals themselves.

Simple three-sided sheds may cost as little as $500 / £298, while full-scale breeding barns with state-of-the art equipment can run to $100,000 / £59,545.

Purely as a hypothetical, a start-up budget for someone who already owns lands might look like this:

- One female alpaca (pregnant) and one young female. (Alpacas should always be kept in pairs.)

  $15,000-$20,000 / £8932-£11,909

- Insurance on the animals. (Note that additional insurance may be required for farm structures and equipment.)

  $500-$600 / £298-£357

- Ancillary equipment, including, but not limited to: halters, shears, toenail clippers, feeders, ropes, etc.

  $500-$750 / £298-£447

- Small barn/shelter and pasture fencing.

  $25,000-$30,000 / £14,886-£17,864

- Feed for one year including hay and pellets.

  $300-$400 / £179-£238

- Veterinary expenses and cash reserve.

  $1,000-$1,500 / £596-£893

Total Estimated Costs: $42,300 - $53,250 / £25,190-£31,708

Estimation of start-up costs is highly case specific, but these figures are a means to jumpstart your thinking about entering the alpaca industry at whatever level feels comfortable for you.

Financing Alpacas

Due to the high cost of alpacas, many breeding operations will finance the purchase of foundation animals for "newbies" just entering the field.

This process works much like buying a car. Buyers put down a percentage of the purchase price, and make monthly payments over a period of years. This is a viable way to "get your foot in the door."

Not all breeding operations offer financing, but it is possible to locate this kind of offer with diligence and research.

Please refer to the breeder directory at the back of this book for lists of alpaca farms in the United States, Canada, and the United Kingdom.

Group Ownership

Another common method of ownership is for a group of investors to pool their resources for the purchase of a breeding pair of alpacas or for a herdsire.

This arrangement eases the matter of financing, but raises the question of where the animal will be housed, and how the husbandry needs will be accomplished.

Ideally, at least one member of the group should own land and have the facilities to keep the animals, with the group sharing the maintenance costs.

Chapter 2 - Understanding Alpaca Fiber

Unless you are acquiring alpacas purely for the pleasure of owning the animals, you will, in some way, be working with alpaca fiber.

In building a foundation herd, alpaca breeders must understand the basic anatomy of the fiber their animals produce, as well as the differences between Huacaya and Suri fiber.

The Structure of Alpaca Fiber

There are three structural sections present in all alpaca fiber:

- cuticle or scale
- cortical cells
- intracellular binder

Overall, the fiber is composed of a protein called keratin, but the fiber's "build" is based on a complex series of cells, each with a unique function.

Cuticle Cells

The cuticle cells are flattened and hard. They do not fit evenly together, and have edges that protrude from the shaft of the fiber creating a serrated edge. The cuticle is responsible for the fiber's aesthetic properties including softness and luster.

As a basis for comparison, the cuticle cells of sheep's wool protrude from the shaft about 0.8 micron. Alpaca cuticle cells protrude 0.4 micron and are thus softer than wool.

It is also from the cuticle that alpaca fiber gains its ability to repel water, to felt when washed, and to resist physical as well as chemical wear.

Cortical Cells

The cortical cells are rounded, elongated, and spindle shaped. Basically, they are thick in the middle and taper at the ends to form a point.

These cells are load bearing, giving the fiber its strength, and are also responsible for its superior ability to repel water.

Intracellular Binder

The fiber's intracellular binder is responsible for holding all of these structures together. Think of it as a sort of "cement."

While all alpaca fiber has the same three main components, there are key differences between Huacaya and Suri fiber.

Huacaya and Suri Fiber

On Huacaya fiber, the cuticle cells protrude a little more than on Suri. Internally, Huacaya fiber is similar to sheep's wool and has a bilateral structure. There are two types of cortical cells: orthocortical and paracortical.

These cells grow in bundles adjacent to one another, creating the characteristic crinkle or "crimp" of Huacaya fiber.

Suri fiber has a lower cuticle height and lesser scale frequency, so it feels slippery to the touch and is more lustrous in its sheen.

Each fiber must be processed differently, with Suri being the more difficult to handle. It lacks cohesion, and is thus harder to spin, with more fiber being lost in the process.

Suri fiber is also heavier and bulkier since more of the straight fibers are needed to make yarn. A 218-yard / 199 meter ball of Suri fiber yarn costs approximately $25 / £15, whereas an equal amount of Huacaya costs is around $18 /

£11. (Please note that it is not unusual for the two fibers to be blended in yarn.

Follicles and Fibers

The fibers grow from both primary and secondary follicles. Primary follicles produce fiber with greater diameter and little to no crimp.

They are the remnants of outer "guard hairs" that have been largely bred out of alpaca genetics in an effort to improve the fiber quality.

The secondary follicles produce the down or undercoat. These are the fine soft hairs that are the hallmark of luxury alpaca fiber. The higher the ratio of secondary to primary follicles, the finer, more uniform, and softer the fleece.

Secondary follicles exhibit a subtype called the derived secondary follicle, structures which have their own root and enter the follicle sheath from the side. These are the finest fibers of all and will be present in a high percentage in alpacas that are considered "elite."

A type of coarse fiber known as "medulated" creates problems for producers when it is present in a fleece. Medulated fiber is not acceptable in premium markets. These hairs do not accept dye readily, and create unacceptable variations in finished yarn.

Fiber Facts

- Alpaca fiber contains no lanolin, making it hypoallergenic with an extremely low "prickle" factor as compared to wool.

- Alpaca's yield 87-97% of their fleece as clean, usuable fiber compared to 43-76% for sheep.

- In testing conducted by the Yocum-McCall Testing Laboratories, alpaca wool was shown to be three times as warm as sheep's wool.

- If won in an ambient temperature of 0 F / - 17 C, alpaca will return a comfort range of 50 F / 10 C.

- The approximate tensile strength of alpaca fiber is 50 N/ktex compared to the industry standard for textiles of 30 N/ktex.

- Alpaca fiber is flame resistant and meets the U.S. Consumer Product Safety Commission's Standard as a Class 1 fiber for use in clothing and furnishings.

Evaluating a Fleece

Judging a fleece has as much to do with sensation or "feel," as it does any set criteria. The best fleeces are:

- smooth
- soft
- even
- slippery (Suri)
- bright (luster for Suri)

There are, however, objective measurements that are applied, and alpaca breeders should be familiar with the terminology of this criteria.

- *histogram* - A histogram is a graph of the distribution of mechanical measurements by diameter and statistical frequency in a given sample of fiber.

- *staple length* - A number of factors influence staple length including, nutrition, environment, and genetics. The greatest staple lengths will be present in Suri fiber, which has no crimp. As an animal ages, it produces progressively shorter staple lengths, as do pregnant and lactating dames.

- *micron count* - By definition a micron is 1 millionth of a meter or 100th of a millimeter. As a basis of comparison, a human hair measures 60 microns. Alpaca fiber is graded by micron measurements that fall into these ranges:

  - Royal: less than 20 microns
  - Baby: 21-23 microns
  - Standard: 24-28 microns
  - Adult: 29-32 microns
  - Coarse: 33-35 microns
  - Very Coarse: greater than 35 microns

  - *standard deviation* - A calculation to express the consistency of micron count across a tested sample. The lower the "SD" number, the greater the consistency of the fiber diameter.

  - *co-efficient of variation* - Expressed as a percentage, CV is another method for describing micron evenness in a sample, providing a more accurate comparison between samples than that which can be derived from the standard deviation alone. Fleeces with exceptional fineness have CV values of less than 20%.

  - *comfort factor* - Fibers with a diameter of more than 30 microns will feel prickly against the skin. The comfort factor is arrived at by subtracting the percentage of fibers with a dimeter greater than 30 microns from 100%.

Density

Density, which refers to the number of follicles per area of skin, is the most important quantitative characteristic that can be applied to the evaluation of a fleece. Obviously animals with higher density produce heavier fleeces, but that fiber will also tend to be finer.

As the density of a fleece increases, the diameter of the primary fiber decreases because the secondary follicles force the primary ones to conform due to the closeness of their alignment.

In a dense fleece, you can see all the way to the animal's skin when the fibers are parted. In fleeces with less density, the primary fibers will be 30-40 microns thick and will cross and intertwine with the secondary fibers obscuring the skin.

You can arrive at a visual estimate of how tightly the fibers are packed by parting the fleece and judging how much skin can be seen at the roots. A fleece that is extremely dense will reveal only a thin line of skin. This is a reasonably reliable method of evaluating a fleece, but there are other approaches.

Pushing the fiber down and feeling the amount of resistance is another method often used. Some people will even just grab the fleece to see how it fills the hand. Both can be misleading. Coarse fibers offer more resistance, and also more completely fill the hand.

Regrowth or Staple Length

"Regrowth" or "staple length" are both terms referring to the actual length of the fiber. Together, length and density determine how much the complete fleece will by, which is the industry standard basis for payment.

From one alpaca to the next, the total weight of the fleece may range from 2-12 lbs. / 0.9-5.4 kg or more.

To judge the rate of regrowth, accurate records must be kept of shearing dates. As alpacas age, they will produce less fiber, a diminishing capacity seen especially in females that are reproducing.

Fineness

Fineness is also important to a good-quality fleece. As shown earlier in this chapter in the definition of micron count, the lower the average fiber diameter (AFD), the finer (and thus softer) the fiber.

Many factors affect AFD including age. AFD tends to increase about 2 points per year until an alpaca has reached 4-5 years of age. Diet and hormones are also relevant influences on AFD. Males have coarser fiber in general, but the fleece of a gelded male retains its fineness due to the lower testosterone levels in these animals.

It's important to use histogram results to learn to assess the fineness of fiber by hand since many of the subjective influences on this crucial factor can be greatly misleading.

A tightly crimped fleece, for instance, will often feel coarser than it really is according to an accurate histogram reading.

Crimp

The ripples or waves in fiber are called the "crimp." The more crimp in a fiber, the finer and denser it is generally believed to be.

This is, however, a "rule" with many exceptions. It is true that fiber with a greater degree of crimp is easier to spin.

In judging the quality of a fleece, the uniformity of the crimp across the entire blanket is important. This is more important than the "style" of the crimp, which is described as high or low frequency in terms of number of crimps per inch.

Lock Structure

The tendency of fleece to separate into cylindrical groups is called "lock structure." This quality is much more evident in fleece from a Suri alpaca than from a Huacaya.

The denser and more uniform the fleece, the more pronounced the lock structure. There is also reference to lock style with Suri fleece, which should twice or wave.

The most desirable style is for small ringlets or waves that are highly uniform. This should start very close to the skin, and be consistent throughout the whole fleece.

Goals for Breeding Programs

Alpaca breeders use these characteristics to create breeding goals, culling from their program animals that do not meet the specified set of criteria.

Culling is not a euphemism for "killing," it simply means selling the animals to another breeder or, in the case of beloved pets, having them spayed or neutered.

Examples of points in a set of breeding goals might include:

- Temperament qualities that relate to ease of handling and manageability.

- High fertility rates with easy births, efficient lactation, good maturation rates, and early weaning.

- Excellence of conformation in regard to bone structure and posture as well as the fleece volume and capacity.

- Specific fleece goals including fineness in microns, density, length of staple, uniformity of color, and brightness.

In any breeding program, specific figures relative to fleece quality are still a guide only, since so many factors can influence the numbers. This does not mean, however, that the numbers should be ignored.

When you are purchasing alpacas, the farmer with whom you are dealing should be able to discuss these factors with you. Don't deal with someone, who says that these measurements are unimportant and that the only determination of fleece quality is that made by "hand."

"Hand" is important, especially in the show ring, but the measured figures are the only way to arrive at an accurate and definitive basis of comparison.

Fleece Shows

As fiber animals, alpacas may be shown for their body conformation and overall excellence, but the fleece itself is also shown competitively.

The British Alpaca Society (www.bas-uk.com) on its "Shows and Events" page states that fleece shows routinely have 200-300 entries.

Participation in fleece events adds prestige to the reputation of an alpaca farm, and award winning fleeces command higher prices when sold for textile production.

Consult the home page of your governing alpaca association to determine what kinds of events are held in your area and to learn more about the criteria for entry.

Chapter 3 - Alpaca Husbandry

The first rule of alpaca husbandry is that these herd animals are best kept in pairs. Beyond that, however, these are marvelously adaptable animals, adjusting to a wide range of climates with the correct help from their keepers.

Visit Alpaca Farms

If you are completely new to the alpaca industry and have never lived on or owned a farm or a ranch, there's no denying the fact that you have a lot to learn! Reading this book is a good start, but there is absolutely no substitute for making friends with people in the industry who live and work with alpacas daily.

Either by attending alpaca shows, or reaching out to existing alpaca farmers in your area, it will be greatly to your benefit to cultivate a mentor or mentors. This will allow you to get hands on experience, discuss pros and cons of equipment, shelters, veterinary care, feed — all the daily details of being an alpaca owner!

And you'll have someone to call when something comes up you didn't anticipate or don't know how to handle! Never underestimate the importance of being able to make that phone call. I think you'll find that alpaca people tend to be a pretty clannish and close-knit bunch. You should have no trouble making new friends, and learning the industry from the inside.

Take the following husbandry information as a primer in the basics so you can go in armed with specific questions. I can't stress strongly enough the need to conduct research and prepare your facilities fully BEFORE you purchase your first alpacas.

Purchasing Alpacas

Although there are many criteria that can be applied to the purchase of an alpaca, you will want to consider some of the following general points:

- Choose animals that have a pleasing appearance to the eye, including straight legs and a body shape that is well balanced.

- Over the "blanket" area, which is essentially the area that would be covered by a saddle blanket on a horse and down the sides toward the belly, the fleece should have an even texture.

- You should only be able to see guard hairs on the front "bib" or chest and potentially some on the neck.

- If purchasing a male for breeding purposes, make certain the testicles are equal in size.

- Buy only from reputable breeders with a solid record in the show ring and see references from satisfied clients.

Upon purchase, you should receive some form of "handover" report that includes information like date of birth, injections received, mating dates if applicable, and ultrasound dates if applicable.

All breeding stock should come with veterinary health certificates which will be necessary for insurance purposes. Pregnant females are generally sold with some form of a live birth guarantee and stud males with a fertility guarantee.

Providing Shelter

In the vast majority of cases, providing a three-sided shelter with one open side, usual on the east to southeast side works well for alpacas as it does for many kinds of livestock.

The reasoning on the positioning of the open side is to shelter the animals from cold north winds and rain while giving them easy and unencumbered access. This arrangement can, of course, vary by region.

The construction of the shelter can follow any number of designs appropriate to region and circumstances. If the shelter is to be temporary, stacked straw bales covered by a tarp is an inexpensive approach.

In more permanent circumstances, barns are used, often with heated floors and automatic watering and misting systems, especially in breeding operations.

Space is the most important consideration. Allow 20-30 square feet / 1.86-2.8 m2 per alpaca.

Concrete floors, if possible, have a distinct advantage, especially when drains are incorporated into the design for washing out urine. Dirt and crushed stone or gravel has to be periodically removed.

Typically straw bedding is provided, especially in the cold months, which also must be "mucked" out to keep it clean of accumulated urine.

Neither wood shavings nor sawdust is a good choice for flooring or bedding material for use with alpacas. Both cause debris in the fiber that lowers the quality of the fleece at shearing time.

Methods of Cooling

Alpacas need shade in the summer to escape the heat of the day, but even on the hottest days you'll find these animals lying on their sides taking in the full sun. This can be disconcerting for first-time alpaca owners, but like cats, alpacas love a good sun bath and will stay there until their fleece is almost too hot to touch.

If you opt for a fan in the barn / shelter, be careful to buy one that is specifically designed for agricultural use. The housing must be capable of keeping high levels of dust out of the motor compartment or the unit will be highly susceptible to overheating, which is a significant fire hazard.

Don't shop for a fan in your local department or discount store. Go to a farm store or agricultural catalog. The cost will be greater, but the safety factor is an imperative.

In more arid regions, the use of misters for cooling is quite common, and also highly effective. Other cooling techniques include sprinklers, soaker hoses, and damp patches of sand.

Alpacas should not have access to ponds or deep water, however, as soaking will cause the fleece to rot and break. If the fiber becomes matted from being wet, heat cannot dissipate from the skin and the potential for overheating will increase. Bodies of water can also be a vector for the transmission of disease and parasites.

In the winter, any body of standing water in a pasture is a significant hazard to the alpacas. They can easily slip on the ice, fall through, and drown.

Fencing and Other Protective Measures

The number of fenced areas or pastures you will require for you alpacas depends on whether or not you will be breeding the animals. If so, you will need to plan for 3-4 fenced areas:

- one to house the females
- one to house the males
- one for young males to protect them from aggression
- one for animals just weaned

Under the best possible circumstances you will have enough land to rotate the animals through a series of pastures to prevent too much grazing and to allow the available forage to recuperate.

If you show alpacas, you may also want to have a quarantine pasture for the animals routinely moved on and off the land to guard against the spread of any infectious diseases.

The purpose of fencing with alpacas is primarily for containment and to keep predators out rather than to keep your alpacas in. They are very passive animals and unlike other types of livestock, will not challenge fences.

Material Considerations

In choosing your fencing materials, remember that alpacas have long necks and exhibit a tendency to stick their heads through open spaces. Traditional cattle fencing with net wire on the bottom and 2-3 single strands on the top offers too many enticing gaps for alpaca.

No-climb horse fencing (welded mesh) works well, as do cyclone (chain link) fences. Try to avoid "New Zealand" or high tensile fences made of single strands of wire at spaced horizontal intervals. Never use barbed wire, and use plastic fencing as a temporary expediency only.

If you can afford to do so, consider having a perimeter fence for predator control that is at least 5 feet / 1.5 meters high. Six feet / 1.8 meters is even better.

For predators with a tendency to dig under fences, bury mesh wire to a depth of at least one foot / 0.3 meters. Electrified fencing is also an option on the perimeter.

Use alpaca safe materials for the interior fence at a height of 4-5 feet / 1.2-1.5 meters.

(Note that the use of guard dogs with your alpacas can be an excellent security boost in areas with high predator activity.)

Diet and Nutrition

Alpacas need amazingly little to survive. On their native ranges in Chile, Peru, and Bolivia, they only have access to lush grasses during the rainy season. The rest of the year, they subsist on sparse vegetation only.

In pastures with an adequate cover of natural, non-fertilized grasses, alpacas will graze contentedly and usually thrive. A mix of 4-6 grasses in a pasture is ideal to create a varied foraging environment. Typically used grasses include:

- Brome
- Orchard
- Timothy
- Endophyte free Fescue
- Winter Wheat
- Bluegrass
- Bermuda
- Millet
- Sudan Grass
- Bahia Grass

Typically a two-week rotation schedule (depending on region) is enough to allow grass to replenish itself. Supplement natural grasses with a low-protein grass hay.

On a daily basis, alpacas eat 1.5% to 2% of their body weight by volume, so a 150 lb. / 68 kg alpaca would need to consume roughly 3 lbs. / 1.4 kg. About 60% of that intake should be from grazing.

The remaining 40% of the diet should come from commercially prepared alpaca feed. This ensures that the animals get vitamins and minerals, like selenium, that can't always be obtained from grass and hay alone.

Although there are many options, an exemplar feed is Mazuri Alpaca & Llama Maintenance Diet, which, according to the packaging, is "designed to maintain adult alpacas & llamas in good condition. This product is not designed for growing, gestating or lactating animals or for fiber animals. It's designed to complement grass or legume hay/pasture."

The suggested feeding directions on the product are, "To be fed with free-choice alfalfa or grass hay or pasture. In order to meet NRC recommendations for new world camelids, animals being fed hay should consume this product at a rate of 0.5 lb. pellet {0.22 kg} per 100 lbs. {45 kg} of body weight (BW).

Hydration

Provide your alpacas with a source of clean, fresh water near their shelter at all times. Although these animals do not consume large amounts of water, they will refuse to drink altogether if their water is dirty or stale.

Shearing

Thanks to the Internet, it's quite easy to go to YouTube and watch any number of videos of alpacas being sheared. The process begins with isolating the alpaca in a contained space called a "catch pen" and then taking the animal into a shearing barn with a concrete floor.

This keeps the fleece as free of debris as possible and allows for better clean-up. The animals are laid flat on their side and the fleece is shorn with electric clippers.

Power Shears

These power shears are made up of a hand piece, comb, and cutters. Portable shears with the motor contained in the handle are recommended for alpacas.

Good quality commercial models that can handle the dense alpaca fleece cost $250-$500 / £150-£298.

The adjustable comb attaches to the hand piece. The flat side faces up and away from the animal. Its purpose is to enter and separate the fibers. This piece dulls quickly, and must be replaced after 2 or 3 animals are shorn. Combs cost $15-$35 / £9-£21 each.

The cutters typically have four triangular points. They attach to the hand piece and press firmly against the comb. Cutters dull even faster than combs, at the rate of about three cutters per single comb. Cutters cost $10-$15 / £6-£9.

Shearing Techniques

Although there are many techniques for shearing animals, some unique to the individual, the basic procedure is that the animal is laid on its side and the shearer works from the belly toward the center of the back.

The fleece is taken off as one single piece or "blanket." The preference is to remove the fleece in one unit, cutting as close to the skin as possible.

After one side is completed, the animal is flipped to the other side and the procedure is repeated. It usually takes three people working as a team, one holding the head and shielding the eyes, and two removing the fleece.

(Many alpaca owners far prefer to hire professional shearers to complete this task.)

The hair is left long on the head and legs, so a shorn alpaca looks a bit like it's wearing a wig and old-fashioned ladies' pantalettes!

(While it is certainly possible for a single person to shear an animal,that requires a fair degree of both experience and confidence. Anyone with experience shearing sheep will have no difficulty shearing an alpaca.)

Shearing and Heat Stress

Even if you are keeping alpacas as pets, the animals should be shorn each year. Failure to do so increases the risk of fatal heat stress. At best, a male that has suffered heat stress can be left permanent by sterile. Signs of heat stress include:

- a wobbling gait
- flaring of the nostrils
- open mouth breathing
- a refusal to stand from the kush position

The best method to alleviate heat stress is to remove the fleece from the body and then the neck. Pour isopropyl alcohol over the body and put the alpaca in front of a fan. If no alcohol is available, cool water will help. Do not hose down a fully fleeced alpaca as the fiber will mat and prevent heat release.

A vet should be consulted as the animal may need treatment, including injections of Banamine and B-complex vitamins.

Alpaca Judging

Like all exhibition shows for pedigreed animals or exceptional livestock, alpaca shows are organized by various governing bodies in the industry. These will vary by region, but there are opportunities for enthusiasts of all ages from children to adults to compete with their animals.

In general, alpaca judges look for the following points:

- Quality of movement when walking toward and away.

- Absence of physical anomalies or abnormalities.

- Width of the chest.

- Nature of boning (fine or heavy.)

- Quality of the top line (Strong or frail and humped back.)

- Fullness of the cap.

- Shape of the ears.

- Shape of the head (wedge.)

- Fullness of a Huacaya's cheeks.

- Quality of chin locks in a Suri.

- Quality of the fleece.

After observing the animals and forming an initial impression, judges conduct a hands-on evaluation before placing the class. Although there are set points to take into consideration, alpaca judging is as much an art as a science and there is almost always disagreement with the judges' final decision!

The Matter of Breed Standards

Although there is no one set breed standard for alpacas, the following points are used by the International Alpaca Judging School. They are reproduced here as an example of the criteria used in evaluating the quality of alpaca show animals. Depending on your location in the world, other standards may apply.

(*Source:* "A Comparative Analysis of Alpaca Breed Type and Standards," by Jude Anderson, Maggie Krieger, and Mike Safley, which can be accessed at the Alpaca Library at www.alpacas.com/AlpacaLibrary.)

GENERAL APPEARANCE - HUACAYAS

The ideal Huacaya alpaca has a squared-off appearance with four strong legs. It is a graceful, well-proportioned animal with the neck being two-thirds of the length of the back and the legs matching the length of the neck. It is well covered with fiber from the top of the head to the toes. It has fiber characteristics that differ distinctly to the Suri alpaca.

GENERAL APPEARANCE - SURIS

The ideal Suri alpaca has a squared off elegant appearance with four strong legs. It is a graceful, well-proportioned animal with the neck being two-thirds of the length of the back and the legs matching the length of the neck. It is well covered with fiber from the top of the head to the toes. It has fiber characteristics that differ distinctly to the Huacaya alpaca.

HEAD - HUACAYAS

The head is neatly formed of medium length with a square muzzle. It bears two upright spear-shaped ears between which there is a full fiber topknot or bonnet. The eyes protrude slightly from their sockets and are large and round.

The eyes can be of several shades although 90% of the population have black eyes. The other acceptable color is brown. There are also various shades of blue eyes with or without colored flecks.

The jaws fit together well, with the lower incisors meeting the upper dental pad. The upper lip is centrally divided and mobile to give them more dexterity in gathering food from certain plants.

The nose has two well-defined flaring nostrils. Darker pigmentation to the skin is preferred around the mouth and eyes giving them added protection to ultra-violet light radiation and the environment.

Major Faults:

- Deafness in blue-eyed alpacas with lack of skin pigmentation and white fleece.
- Gopher ears.
- Superior and inferior prognathism.
- Wry face.
- Lump on the side of the face indicative of abscessing in the mouth.
- Eyes: cataracts, entropy, ectropy, blindness.

Minor Faults:

- A straight inside border or banana-type configuration of the ear (indicating llama traits).
- Forward set ears.
- Roman nose (llama tendency).
- Narrow head.
- Muffled face in the adult alpaca. (fiber or hair impeding the alpaca's vision).
- Open-faced. (Lack of fiber coverage over the face.)
- Lack of pigmentation on the lips and around the eyes.
- Retained or persistent deciduous teeth.

HEAD - SURIS

The head is neatly formed of medium length with a square muzzle. Suris have more of a tapering shape to the muzzle. They bear two upright spear-shaped ears between which there is a full fiber topknot or bonnet that falls typically in a fringe over the brow. Suri ears are approximately 2cm longer than Huacaya ears.

The eyes protrude slightly from their sockets and are large and round. The eyes can be of several shades although 90% of the population is black. Brown is also a desirable color. There are also various shades of blue with or without colored flecks.

The jaws fit together well, with the lower incisors meeting the upper dental pad. The upper lip is centrally divided and mobile to give more dexterity for feeding off certain plants.

The nose has two well-defined flaring nostrils. Darker pigmentation to the skin is preferred around the mouth and eyes giving them added protection to ultraviolet irradiation and the environment.

Major Faults:

- Deafness in blue-eyed alpacas with lack of skin pigmentation and white fleece.
- Gopher ears.
- Superior and inferior prognathism.
- Wry face.
- Lump on the side of the face indicative of abscessing in the mouth.
- Eyes: cataracts, entropy, ectropy, blindness.

Minor Faults:

A straight inside border or banana-type configuration of the ear indicating llama traits

- Forward set ears.
- Roman nose (llama tendency).
- Narrow head
- Muffled face in the Suri (fiber or hair impeding the alpaca's vision or retained on the adult face).
- Retained or persistent deciduous teeth
- Open faced with lack of fiber coverage over the face. Lack of pigmentation around the lips and eyes

HEIGHT AND WEIGHT

The height at the withers of the adult alpaca is no less than 85cm (32") and the average weight of an adult alpaca is 60kg (140lbs).

Faults:

- Small sized with less than 85cm (32") measurement at the withers.
  Oversized with llama characteristics.

LEGS

The legs are supported by four two-toed feet, with each toe supporting a long toenail. They should be straight with the joints aligned to a perpendicular plumb line from the hip posteriorly and shoulder anteriorly. The shoulder blade is attached by muscular tissue to the thoracic cage but should move freely as the animal strides. A leathery padded membrane, which lessens the impact on the environment where they tread, protects the feet.

Major Faults:

- Excessive angular limb deformity.
- Subluxing patellae.

Minor Faults:

Front Legs:

- Knocked knees.
- Calf knees/cocked pasterns.
- Bucked knees/dropped pasterns.
- Medially or laterally deviated pasterns.
- Splay legs.

Rear Legs:

- Cow hocks.
- Sickle hocks.
- Bowlegs.
- Cocked pasterns.
- Dropped pasterns.
- Poorly maintained toenails.

BODY

The neck of the alpaca is straight and upright and blends

smoothly into the back, which is normally very slightly rounded in the Huacaya.

The rear of the alpaca has a tucked-in tail appearance that is due to the angulation of its pelvis being more vertical than the llama, sitting at about 60 degrees from the horizontal.

The resting position of the tail is such that it lies close to the body, covering the genitalia. The tail is raised away from the body during defecation and urination and for expression of temperament and mood. This gives the tail a noticeably lower set than that of the llama.

The chest should have depth to allow adequate capacity for air exchange.

Major Faults:

- Lateral deviations of the spine.
- Herniated umbilicus.

Minor Faults:

- Roach back.
- Sagging back
- U neck
- Lateral deviations of the neck
- Disproportionate length of neck (too long or too short).
- Deviations of the tail, broken tail.

GAIT

A free-flowing stride is characteristic of the alpaca. Its normal slow speed gait is a stable four-point gait where each foot is moved and planted separately. At a faster speed the alpaca has a pacing gait which is two-point, where the two feet on either side are moved together.

Major Faults:

- Excessive angular limb deformity causing excessively abnormal movement.
- Subluxing patellae causing abnormal rear gait.

Minor Faults:

- Joints tracking medially or laterally to the vertical plum line.
- Gaits associated with angular limb deformity such as winging, arcing, rope walking and throwing out of the front limbs where there is rotation at the joints of the front limb.

GENITALIA (Female)

The genitalia of the female is protected internally and therefore not visible from the outside. However, the vaginal opening should be well covered by the tail, should not be too small and should be situated in a vertical rather than a horizontal plane.

Major Faults:

- Too small of a vaginal opening.
- Hemaphroditism.
- Lack of any part of the reproductive system.

Minor Faults:

- Horizontally situated pelvic floor.
- Tipped up clitoris.

GENITALIA (Male)

The most visible part of the male genitalia is the testicles that are situated and protected underneath the tail. The scrotum is well attached, relatively small and carries the testicles, which are relatively even in size.

The penis is also an external organ, which is situated under the belly between the rear legs. The normal size of fully developed testicles is: 4cm in length, 2.5cm in width in the adult male alpaca.

Major Faults:

- Hermaphroditism.
- Ectopic testicles (these testes are located outside the abdominal cavity under the skin, sometimes migrating down the leg).
- Cryptorchidism of the testicles/unilateral or bilateral (the lack of one or more testicles in the scrotum). Too soft or too hard testicular consistency.

- Cystic testicles.
- Unilateral or bilateral hypoplasia of the testes (one or both testicles of abnormally small size for the age)

COLOR - HUACAYA AND SURI

Huacaya fleece comes in varying shades of color: white, fawn, brown, gray, rose-gray, and black. The ideal alpaca should have a uniform solid color throughout the entire fleece. However they can be any combination of the above.

COLOR - SURI

Suri fleece comes in varying shades of color white, fawn, brown, gray, rose-gray, and black. The ideal Suri alpaca should have a uniform solid color throughout the entire fleece. However they can be any combination of the above.

FLEECE - HUACAYA

Huacaya alpacas produce a fine soft fiber that grows perpendicular to the skin. In the ideal Huacaya alpaca there is marked crimp formation as the fiber grows out of the skin. The hair follicles are situated close together in the skin, giving density to the fleece with groups of fibers bunching together to form defined staples. The following fiber characteristics are applicable to Huacaya fiber:

1. *Fineness* - this is the thickness of the fiber that is measured in microns. The finest fiber on the alpaca is found in the blanket area, however it is desirable to have fine fiber on the neck, belly, legs and topknot.

Fineness is important for both commercial processor and the fiber grower since premium prices are paid for fine fiber and fine fiber translates into fine end products. Crimp is also related to fineness and it is desirable too to have a high number of waves per cm. or inch.

2. *Density* - is the number of fibers per square measurement of skin. Density is associated with fleece weight since the more fibers per square unit measurement, the more fleece will be grown and the heavier the fleece. A dense crimped fleece also acts as a barrier to dirt and weather.

3. *Character* -defined as strong crimp definition and staple formation.

4. *Length of staple* - is a very important factor in the amount of fleece shorn from the Huacaya alpaca. The more rapidly the length of staple that is grown the more weight of fleece there will be.

5. *Brightness* - is the amount of light that reflects from the fiber and is seen in the Huacaya. A brilliant appearance of the fleece is desirable.

6. *Medulated fiber* - is the coarse-microned fiber that grows in the lesser quality areas of the alpaca. Lack of medulated fiber in the prime or blanket area is desirable.

7. **Uniformity of micron** - processors require fleece of minimum variation in fiber diameter, therefore uniformity in fiber diameter is desirable across the blanket area of the alpaca. This also helps to eliminate fleece tenderness (fleece breakage) and prickle effect in the end product.

Faults:

- Open fleece with no density
- Harsh handle
- Short staple length
- Guard hair in the blanket
- Lack of overall coverage
- Tenderness and stress breaks
- Felting and cotting

FLEECE - SURI

The primary characteristics of the Suri fleece are its lock structure, high luster, silky handle and long staple length. The fleece falls close to the body, moves freely, and gives the Suri a flat-sided, lustrous appearance.

The locks can have a penciled ringlet formation, curling to the left or right, or a wave structure that forms from the skin of the alpaca. The fleece locking should begin from the forelock and continue uniformly down the neck, across the blanket and through the legs. The following fiber characteristics are applicable to Suri fiber:

1. *Fineness* - this is the thickness of the fiber, which is measured in microns. The finest fiber on the alpaca is found in the blanket area, however it is desirable to have fine fiber on the neck, belly, legs and topknot. Fineness is important for both commercial processor and the fiber grower since premium prices are paid for fine fiber and fine fiber translated into fine end products.

2. *Density* - is the number of fibers per square measurement of skin. Density is associated with fleece weight since the more fibers per square unit measurement, the more fleece will be grown and the heavier the fleece.

3. *Lock Structure* - in the Suri lock structure is very important. The fibers group together to form ringlet type locks that turn to the right or to the left. Ideally, the lock should form a ringlet from the skin. However, it is common to find a lock structure that starts at the skin as a flat wave formation then continues out down the side of the alpaca in a ringlet.

4. *Luster* - is the sheen or shine that reflects from the fleece. This is a highly desirable trait in the Suri fleece and translates in the end product. The smooth flat structure of the outside cuticular layer of the individual fibers is responsible for this trait.

5. *Length of staple* - is a very important factor in the amount of fleece shorn from the Suri alpaca. The

more length of staple that is grown the more weight of fleece there will be. A Suri will grow 60% longer fleece than Huacaya in one year's growth.

6. *Medulation* - there should be little or no evidence of medulated fibers in the fleece.

Faults:

- Open fleece lacking lock definition.
- Lack of density.
- Crimp.
- Harsh handle.
- Short staple length.
- Guard hair.
- Lack of overall coverage.
- Tenderness and stress breaks.
- Felting and cotting.

Chapter 4 - Alpaca Health and Breeding

Alpacas are highly adaptable animals. They have a hardy constitution cultivated over millennia living in the high, hostile Andes Mountains. It is rare for these animals to need more in the way of standard veterinary care than:

- castration
- worming
- annual inoculations
- Vitamin D supplementation

- toenail clipping

Because shearing of pets is a way to avoid heat stress it is technically a "healthcare" procedure, but obviously if the alpacas are kept as fiber producers, shearing occurs annually for reasons of profit.

Preferably BEFORE you purchase your alpacas, you should locate a qualified veterinarian who will agree to care for your animals. Under the best circumstances, you will find a doctor with previous experience treating the animals.

Most "large animal" veterinarians who have treated sheep and goats can work with you to care for your alpacas, but it is imperative that the vet is agreeable to learning about the species and to consulting with other veterinarians who have expert knowledge in alpaca care.

It is to your benefit and to that of your animals that you cultivate a good working knowledge of alpaca healthcare needs. This will not only help you to make good decisions about the welfare of your animals, but also to control costs in this area of alpaca care.

At the very least, developing a working understanding of alpaca healthcare will help you to ask better questions.

I do not personally like to move forward with a healthcare procedure or treatment with any kind of animal until I am comfortable that I understand what is being done and why.

This is very much like the concept of "informed consent" in human healthcare. To truly be informed about the care your animals are receiving, you need to understand both the potential benefits *and* complications.

Body Scoring

As a basic evaluation of overall alpaca health, body scoring is a simple and fast method to determine the general state of the animal's wellbeing.

- Hold or stand by the side of the alpaca.

- Place your hand flat on the animal's back about 6" / 15 cm behind the withers.

- Put your palm on the spine and your thumb on the ribs to one side of the backbone with your fingers to the other side. Press down firmly until you can feel the spine and ribs.

- In healthy animals, you will feel a smooth line of flesh running from the spine to the ribs that is not indented (concave) or rounded (convex). Your thumb and forefingers will be in the shape of a "V."

- If the spine protrudes upward and into your hand and the ribs are concave, the alpaca is overly thin.

- If your thumb and fingers are relatively parallel and vertical to the animal, the alpaca is dangerously thin.

- If your hand opens wide with bulging flesh between the spine and ribs, the alpaca is overweight. If your hand is horizontal and essentially flat, the animal is obese.

This simple evaluation can be performed daily in the morning or evening, and is easily done while you are feeding the alpacas.

Castration or Gelding

Many alpaca breeders castrate or "geld" males they do not intend to use for breeding purposes or that will be sold as pets when the animals are less than a year old. It's estimated that more than 80% of male alpacas are gelded, with only the top 10% used in breeding programs.

If your purchase a young male alpaca (18-24 months), the animal will likely already have been gelded, but be sure to ask. Since alpacas are herd animals that should not live alone, an intact male will have aggression issues due to the testosterone in his system.

At around three years of age, male alpacas develop "fighting teeth." These teeth grow into the lower jaw between the incisors and molars. They are very sharp and grow continuously, requiring annual trimming. Fighting males can seriously injure each other if this chore is not performed.

As an example of just how vicious an alpaca fight can be, and how seriously these animals take the matter of herd dominance, in the wild, a dominant male alpaca will sometimes use his fighting teeth at the end of a particularly violent struggle to castrate his opponent.

Worming

Like all livestock, alpacas become infested with parasites when they consume the eggs or larvae while grazing. If your local veterinarian has no previous experience with alpacas, the best course of action is to take a fecal sample to determine the correct deworming agent to use.

If no other animals are being kept on the land, alpacas can be wormed twice a year, typically in May and November. When other livestock, especially sheep, are kept in the same field, you may need to worm more often.

As with sheep, worming is accomplished via "drenching," which is the administration of a liquid dewormer orally. Care must be taken, however, that the parasites present don't become resistant to the deworming agent.

Again, advice should be sought from your local veterinarian and other, more experienced alpaca farmers in your region about the frequency with which drenching should be administered and in what dosage.

But generally worm at 6 month intervals eg May and November. But if you have other livestock you'll need to worm more often (consult your vet)

I feel it's important to rotate worming products to help avoid building up resistance.

(Please note that if you are keeping alpacas in an area with a native population of whitetail deer, your animals will require monthly injections as protection against meningeal worms.)

Shearing

Obviously animals that are raised primarily for their fiber will be sheared on an annual basis. If you are keeping alpacas purely as pets, Huacaya alpacas must be sheared annually to guard against heat stress.

Suri alpacas can be sheared every other year. Shearing should be timed so that the animals have regrown at least 1 inch / 12.54 cm of fleece before the weather turns cold.

(Please see the previous chapter on alpaca husbandry for a description of the shearing process.)

Teeth

The front teeth shouldn't protrude from the upper jaw as this can make grazing difficult. You'll need to trim the teeth back. Your vet is probably best as they can do this for you with clippers or dental wire. You may wish to consult a horse dentist who generally have a good understanding of this problem and can help reshape the whole jaw if required.

If you have a group of males you should trim the fighting teeth back to help stop them damaging other alpacas whilst they sort the pecking order out!

Feet

The toenails should not grow longer than the edge of the toe. If they do simply trim them back with sheep foot rot shears. Make sure you clear the dirt out so you can clearly see where the toe flesh is. The foot should be able to stand flat on the ground without twisting.

Vitamin D Supplementation

During the winter months, alpacas may require Vitamin D supplementation. Darker animals with dense fleece are even more susceptible to Vitamin D deficiency and to developing rickets. Again, however, the need for Vitamin D supplementation varies greatly by region and climate as well as by age of the animal.

Overdosing with Vitamin D can lead to organ failure. The supplement can be given as an injection or orally, but both forms should not be used at the same time.

Most vets are in agreement that the injectable Vitamin D is more readily absorbed and can be administered once every 60 days. The oral form is often given every 2 weeks.

A veterinarian should always be consulted before Vitamin D or any other kind of supplementation is used with your alpacas. Correct dosing with these products is essential.

Vaccination Protocols by Region

Working with your local veterinarian to determine the correct inoculations for your alpacas is critical. Disease and parasites VARY GREATLY by location and climate. These factors MUST be considered in developing a vaccination protocol for your animals.

In the UK, for instance, alpacas are vulnerable to the Bluetongue serotype viruses that are active on the continent as well as to local parasites like strongyle worms. Some areas of Britain also see a high prevalence of liver flukes.

In Britain, the recommended clostridial vaccinations are:

- Lambivac
- Covexin 10
- Heptavac-P Plus (includes Pasteurella)
- Ovivac or Ovivac-P

Heptavac is used to protect against:

- Lamb Dysentery
- Struck
- Pulpy Kidney
- Braxy
- Blackleg
- Tetanus
- Black Disease

Tetanus can develop all too easily from a simple cut. For this reason, annual CDT injections that protect against both tetanus and clostridial diseases are recommended.

Depending on your location, assume that your alpacas are vulnerable to all the diseases and parasites from which sheep can suffer.

In the U.S., recommended vaccinations include, but are not limited to:

- IMRAB 3 (rabies)
- CDT (tetanus and clostridium)
- leptospirosis

These and other vaccinations used with alpacas in the U.S. are considered "off label" and have not been approved for use with these animals by the U.S. Department of Agriculture. Individual vets may have different opinions about which drugs work best.

(Please note that this information is provided as the basis for a conversation with your veterinarian about the vaccinations your animals will require and is not intended to be taken as a set protocol. I cannot stress strongly enough that required inoculations vary by region and should be dispensed according to expert veterinary advice.)

Toenail Clipping

Trimming your alpaca's toenails will be as easy or as hard as the animal decides to make it. This is a job that absolutely demands patience, and a willingness to concede defeat in any single session and finish the job on another day.

Standing on three legs while one foot is being held up for the trimming is an act of supreme trust on the part of the alpaca. Remember that you are dealing with a prey animal whose primary instinct when he feels threatened is to run.

In the wild, on the rocky slopes of their native South American mountainous range, alpacas keep their hooves worn down naturally. When they are pastured on soft ground, however, trimming is essential.

Animals with light colored toenails will need even more frequent trimmings as dark nails are harder and grow much slower. In some instance dark nails will only need to be trimmed annually at the time the animal is being sheared.

If the nails are not trimmed, they may cause the toe to twist painfully and pinch the pad. Long nails ultimately break off, leading the animal to go lame. It is even possible for a nail to overgrow to the point that it perforates the pad and causes a painful wound.

If you look at the underside of an alpaca's foot, you will see two toes and the soft pad. There are two nails.

- Cradle the foot in your hand, with the underside up and visible.

- Use a pair of garden pruning shears.*

- Carefully trim the nails until they sit level with the bottom of the pad.

*There are many kind of clippers that will work well including those specifically designed to be used with sheep and goats. The important thing is that the implement be comfortable in your hand for maximum control and sharp enough to accomplish the trimming quickly and efficiently.

Conditions Common to Alpacas

All of the following conditions are common to alpacas, but like all health matters concerning these animals, are greatly affected by location. Also, this is not an all-inclusive list, but rather an overview of conditions commonly associated with alpacas.

Rickets or Vitamin D Deficiency

Alpacas of less than 2 years of age as well as females who are pregnant or nursing can be susceptible to rickets as a result of Vitamin D deficiency. This is also true of animals with especially thick fleece.

During the winter months the lower levels of sunlight cause an abnormal ratio of calcium to phosphate, which affects bone growth. Demineralization of the long bones, called osteomalacia, accompanies rickets and presents with a painful series of symptoms including:

- a hunched posture
- obvious discomfort when moving
- walking slowly with legs splayed
- the appearance of leaning backward while walking

An affected animal will lag behind the rest of the flock and spend most of its time in the kush or resting position with the legs tucked under the body.

Treatment for rickets includes injections of Vitamin D and phosphorous supplements, but care must be taken not to overdose the alpaca with toxic levels. Consultation with a knowledgeable veterinarian is essential.

Tuberculosis

Alpacas have little if any resistance to tuberculosis and are extremely vulnerable to the disease. This susceptibility is complicated by the fact that there is no reliable TB test that can be used. The skin test used widely with cattle detects only about 20% of cases in alpacas and the blood test is equally unreliable and often gives false positives.

Signs of tuberculosis in alpacas include:

- lethargy
- self-isolation from the group
- weight loss (often sudden)
- coughing

If the disease develops in the thoracic and abdominal cavities there may be no visible signs. TB often progresses so rapidly that the animals are simply discovered dead in the pasture for no apparent reason.

Under these circumstances, it's important that a vet examine the body to determine if TB is present. Interestingly, however, alpacas do not then seem to transmit the disease among themselves, but rather to contract it from other infected livestock, usually cattle.

TB can also be transmitted from wildlife to alpacas. In the UK, for instance, tuberculosis is present in Shropshire badgers.

These animals come into the pasture to forage on animal droppings and can infect alpacas grazing in the area. Badgers can only be kept out of fields by burying wire three feet / 1 meter into the ground on the fenced perimeter.

Check with your local alpaca association to see if there is a program for testing and reporting cases of tuberculosis.

IMPORTANT NOTE: Tuberculosis is a zoonotic disease and can be transmitted from animal to humans.

Dietary Poisoning

All grazing animals can be inadvertently poisoned by plants with which they come into contact in the pasture. This is also true of garden plants to which alpacas may be exposed on smaller farms adjacent to planted yards.

Plants known to be poisonous to alpacas include, but are not limited to:

- nightshade
- rhododendrons
- lilies
- azaleas
- solanums
- inkweed
- foxglove
- oleander
- hemlock
- willow weed
- tutu (toot)
- ragwort
- ngaio
- datura
- mallow
- Jerusalem cherry

- yew
- buttercup
- laurel
- irises
- macrocarpa
- bracken fern
- helebores
- daffodils

If you suspect that your alpaca has ingested a poisonous plant, immediately seek the assistance of a qualified veterinarian.

Facial Eczema

Facial eczema in livestock is caused by a mycotoxin in the pasture. The spores of various types of fungus contain sporidesmin, a toxic chemical that damages the liver, preventing the normal breakdown of metabolic and digestive toxins in the bloodstream.

As these toxins build up, the compounds that leach into the skin react to sunlight. Clinical symptoms include:

- skin irritation
- crusting and oozing of the ears and nose
- decreased growth rates in young animals
- spontaneous abortions

Since alpacas hide signs of illness, liver disease is often not discovered until the animal has succumbed to a sudden death and a liver biopsy is performed.

Ryegrass Staggers

In areas where perennial ryegrass is popular, alpacas can be exposed to the endophyte fungus *Acremonium lilii*. When ingested, the mycotixns produced by the fungus attack the brain and central nervous system causing "ryegrass staggers."

Symptoms of the condition include tremors of the head and neck and an unstable gait (ataxia). If left untreated, the alpaca will collapse and die.

Susceptibility seems to have a genetic component, and ryegrass staggers are seen more frequently in the summer and autumn. Hay cut from an infected pasture remains toxic, however, so animals can be symptomatic at any time.

If treated at an early stage with Mycosorb or Biomass in combination with good quality lucerne hay and fresh water, the alpaca will recover within a few weeks.

Bovine Viral Diarrhea (BVD)

Bovine viral diarrhea is a disease that has affected alpacas in North America since 2001. It is acute and short term, and if the alpacas are healthy, their robust immune system is usually capable of eradicating the illness quickly.

However, if the virus is contracted by a pregnant female, the chances that she will abort the fetus are high. If the baby lives, it may well be infected with BVD persistently and be a carrier of the virus.

Overview of Alpaca Breeding

Obviously the subject of alpaca breeding can be addressed from the perspective of a business, which I did in the first chapter, or in terms of the "mechanics," which I will discuss now.

Mating

Alpaca females only ovulate during the act of mating, conceiving shortly afterwards, which makes artificial insemination difficult. This type of reproduction is called "induced ovulation" and is stimulated both by the motion of the male and by the "orgling" noise he makes.

A young female alpaca is ready to be placed with a stud male when she is 14 months of age, or when she has attained 60 % of her mother's weight.

Male alpacas are ready to breed when they are 2-3 years of age, although some are capable of mating as young as nine months.

The selection of the stud male is the critical decision in creating a mating pair as the male has the greatest influence on the quality of the offspring. This why only about 10% of alpaca males are left intact to be herdsires. Most males are gelded and kept for their fiber or sold as pets.

During copulation, the female sits for the male in the kush position. He mounts her from behind. (Females that are not receptive refuse to sit and will often spit at the males to reject their advances.) Because male alpacas are dribble ejaculators, mating may last as long as 45 minutes.

Do not be alarmed if there is evidence of blood at the completion of mating, especially if the female alpaca is being bred for the first time.

Gestation and Birth

The average period of gestation is 345 days, but varies from 330 to 370 days. The best time of year for the young, called cria, to be born is late spring into early summer. If, however, the correct shelters and facilities are in place, many breeders plan births from March through October.

Typically alpaca births occur during the middle of the day and are free of any trouble. The cria should be 12-20 pounds / 5.4-9 kg at birth and within 2-3 hours will be on their feet and nursing. Mothers are quite protective and will not wean their babies for 5-6 months.

Planning Matings

Even if a female is nursing her cria, she can still be re-mated 2-6 weeks after giving birth. This means that in planned breeding programs, maintaining multiple pastures to separate the breeding animals is essential.

For this reason, pen mating is often the best method. This requires a space 10-14 feet / 3.04-4.26 meters gated on one side. In good weather, this is easily created with portable fencing in a pasture. The pen sides should be 4-5 feet / 1.2-1.5 meters high. In pasture mating, it's best to choose an area new to both animals.

Confirming Ovulation

Seven days after the initial mating, the pair should be re-introduced. If the female refuses to sit for the male, the chances are quite good that she has indeed ovulated and conceived. Repeat this process at days 14, 21, and 28 to confirm the outcome. (Large breeding operations will test via ultrasound.)

Chapter 5 - Frequently Asked Questions

While it is necessary to read the entire text to get an understanding of both alpacas and the alpaca industry, the following are some of the questions I am asked most frequently about the animals and their fiber.

What is an alpaca?

Alpacas are "camelids" indigenous to South America. They are closely related to llamas, and less so to Asian and African camels.

These docile herbivores produce exceptionally high quality textile fiber, and their cultivation for this purpose on a worldwide basis has been steadily growing since the 1980s.

What is a Suri alpaca?

Suri alpacas have unique fiber characteristics different from Huacaya alpacas. Their fiber is extremely fine and soft, hanging in long, lustrous locks with no crimping. It grows parallel to the body and may be either flat or twisted.

Suri fiber is highly sought after in the fashion industry and by spinners and weavers. It is very like cashmere, with a silky touch. Though lightweight, it is exceptionally warm and durable and can easily be blended with wool and silk. This type of alpaca is very rare, however. As an example, sufis make up only about 10 % of all the alpacas present in North America.

What is a Huacaya alpaca?

The fiber of a Huacaya alpaca is wooly in appearance, dense, and crimped. The resulting look on the animal itself is that of a huggable teddy bear. In North America, about 90% of all alpacas are Huacayas. Their fiber is fine and highly prized.

What is the difference between alpacas and llamas?

One of the easiest ways for the newcomer to this industry to tell an alpaca and a llama apart is to look at the ears. Alpacas are physically smaller than llamas, but their ears are straight while llamas have curved ears that look rather like bananas.

In terms of the fiber, the coat of a llama is interspersed with coarser guard hairs that must be removed during processing, while alpaca fiber is very fine and soft.

Behaviorally, llamas are much more aggressive and are often used as guard animals for grazing herds of sheep and even alpacas themselves.

Do alpacas make good pets?

Although alpacas are docile and social, their primary reaction to everything in their environment stems from their existence as prey animals. In general, they are wary of humans.

They do not like to be grabbed, and there are areas of the body (feet, lower legs, and abdomen) where they do not like to be touched. If handled well, they will interact peacefully with humans, but they won't come when they are called like dogs or cats, nor do they really like much in the way of petting. Some individuals are however, more affectionate than others.

Are alpacas smart?

Yes, alpacas are highly intelligent and adaptable. They have inquisitive and curious natures, but do follow the dominant herd members. This does not, however, prevent them from learning new tasks quickly and being cooperative with their handlers. Although not necessarily making good "pets," per se, they do seem to enjoy their interaction with humans and some individuals can be quite affectionate.

Do alpacas spit like camels?

Alpacas do have the ability to bring up acidic solid material from their stomachs, which they will spit at one another, but rarely at humans. This behavior is most often seen during feeding times or other situations where dominance comes into play. It is not a harmful behavior, and doesn't escalate into more serious confrontations.

What do alpacas eat?

Alpacas are grazing herbivores, existing primarily on a diet of grasses. They do not, however, pull the grass up by the roots, so the pastures on which they are placed will renew if the herd is rotated regularly.

They do best on low protein grasses. If fed higher protein forage like alfalfa or clover, alpacas may experience health problems and the quality of their fleece will suffer.

What kind of shelter do alpacas need?

Sheltering alpacas from the elements is strictly determined by location. In areas where winter temperatures fall below freezing and heavy snowfall is the norm, a three-sided shelter will protect the animals from the elements.

In areas with high summer heat and a lot of humidity, the same three-sided type shelter can be used to provide shade, but an additional means of cooling (like the use of sprinklers in the pasture) will be needed.

Put up a permitter fence in keeping with the degree of security necessary to protect the alpacas from the types of predators indigenous to your area.

How large a pasture will I need for my alpacas?

The amount of pasture you will need for your alpacas will depend on the ability of the land to produce quality grass in relation to climate and precipitation. In general, however, an acre / 0.4 hectare of good quality pasture will support approximately 5-10 alpacas.

What is the lifespan of an alpaca?

The expected lifespan of an alpaca is 15-20 years.

How many offspring do alpacas have at one time?

Alpaca females deliver one offspring at a time. The young are called "cria." It is very rare for an alpaca to deliver twins and when this does occur, the babies rarely survive.

What is the length of gestation period for alpacas?

Alpaca females carry their offspring for 11.5 months, but this may vary by 30 days or more. Alpaca females are induced ovulators. The act of breeding stimulates their ovulation, which allows breeders to time deliveries for the most favorable seasons of the year.

Why is alpaca fiber considered to be so special?

Alpaca fiber is an extremely fine and rare specialty fiber, exhibiting a quality exceeding that of cashmere. The hollow fiber has superior insulating abilities, and is five times warmer than wool but lightweight and free of the coarse guard hairs seen in llama fiber.

The entire fleece "blanket" removed from an alpaca is suitable for use in the production of garments, and there are at least 22 naturally occurring colors. (The recognized number of colors varies by country.) As an added plus, alpaca does not cause the itching associated with garments made from sheep's wool.

Chapter 6 – Breeder Directory

Please note that the following breeders and farms were extant in mid-2014 at the time of this writing. Due to the ever changing nature of the Internet, no guarantee can be made that these addresses will be valid in the future.

Please refer to the Relevant Websites section for various alpaca professional organization that maintain continuously updated breeder directories.

The following list is a representative sampling and is not intended as a comprehensive resource.

U.S. Alpaca Farms

Alabama

Timberlake Farms Alpacas
alpacasalabama.com

Arizona

AlAnn Ranch Alpacas
www.alannranch.com

Cienega Creek Farm
www.cienegacreek.com

Gemini Star Alpacas Farm
alpacasarizona.com

High Country Alpacas, LLC - Suri Alpacas
www.highcountryalpacas.com

Peaceful Prairie Ranch
www.peacefulprairie.com

Arkansas

Ozark Mountain Alpacas
www.ozarkmountainalpacas.com

California

101 Alpacas Ranch
www.101alpacas.com

California Alpacas
www.california-alpacas.com

Alpacas at West Ranch
www.alpacasatwestranch.com

Alpacas By The Sea
www.alpacasbythesea.com

Dancing Moon Alpacas
www.dancingmoonalpacas.com

Friendly Farm Alpacas
friendlyfarmalpacas.com

Lassen View Alpacas
www.lvalpacas.com

Luv R Pacas
www.luvrpacas.com

Rancho Keleje Alpacas
www.ranchokelejealpacas.com

Rockstar Alpacas
www.rockstaralpacas.com

Shiloh Springs Ranch
shilohspringsranch.com

Spring Oaks Alpacas
www.springoaksalpacas.com

Sunny Acres Alpacas
www.sunnyacresalpacas.com

XC Alpacas Sheep & Goats
www.xcalpacassheepgoats.com

Colorado

ABC Farm
abcfarmcolo.com

AbraCadabra Alpacas
www.abra-cadabraalpacas.com

Alpaca Annuals
www.alpacaannuals.com

Alpacamundo
alpacamundo.com

Alpacas at Whiskey Creek
www.whiskeycreekranchcolo.com

Alpacas de La Mancha
www.alpacasdelamancha.com

Alpacas of Vista Hermosa
www.alpacasofvistahermosa.com

Argenta Alpacas
www.argentaalpacas.com

Aristocrat Alpacas
www.aristocratalpacas.com

Bar 3 Suri Alpaca Ranch
www.bar3Suriranch.com

Big Hat Ranch
bighatalpacas.com

Black Forest Alpacas
www.blackforestalpacas.com

Blazing Star Ranch
www.blazingstarranch.com

Bliss Ranch Alpacas and Ultrafine Merinos
www.blissranch.com

C2 Alpacas
c2alpacas.com

Casa Lavanda Alpacas
www.casalavandaalpacas.com

Corona Trail Alpacas, LLC
www.coronatrailalpacas.com
Delphi Alpacas
delphialpacas.com

Dry Creek Alpacas
www.drycreekpacas.com

Falkor Ranch
www.falkorranch.com

Green Pastures Alpaca Ranch
www.greenpasturesalpacaranch.com

Honeycomb Alpacas
www.honeycombalpacas.com

Lazy Daze Ranch Alpacas
www.lazydazeranch.com

Linden Hills Alpaca Farm, LLC
lindenhills.openherd.com

Little Circle Farm Alpacas
littlefarmalpacas.com

Little Lost Creek Alpaca Farm
www.littlelostcreekalpacafarm.com

Marquita Ranch Alpacas
marquitaranch.com

NeverSummer Alpacas, LLC
www.neversummeralpacas.com
Peak Ranch Alpaca
www.peakranchalpacas.com

Pleasant Journey Alpacas
www.pleasantjourneyalpacas.com

Powers Alpacas LLC
www.powersalpacas.com

Prairie Moon Alpacas
prairiemoonalpacas.com

Rancho Alamogordo
www.ranchoalamogordoalpacas.com

Silken Suri Alpaca Ranch
www.silkenSuri.com

SunCrest Orchard Alpacas, LLC
www.suncrestorchardalpacas.net

Xanadu Farm Alpacas
www.xanadualpacas.com

Connecticut

Alpaca Obsession
www.alpacaobsession.com

Independence Farm
www.ifalpacas.com

Morning Beckons Farm
www.morningbeckonsfarm.com

Pine Hill Alpaca Farm
www.pinehillalpacafarm.com

Six Paca Farm
www.sixpaca.com

Still Meadow Alpaca Farm
www.stillmeadowalpacasfarm.com

Stone Bridge Alpacas
www.stonebridgealpacas.com

Florida

Beloveds Farm
www.belovedsfarmalpacas.com

Funny Farm Alpacas
www.funnyfarmalpacas.com

LunaSea Alpaca Farm
www.lunaseaalpacafarm.com

STARanch Alpacas
www.staranchalpacas.com

Sun Spiced Alpacas
www.sunspicedalpacas.com

Sweet Blossom Alpaca Farm
sweetblossomalpacas.com

TMMA Farms
www.tmmafarms.com

Woodland Hills Alpacas
www.woodlandhillsalpacas.com

Georgia

Alpacas4u2cfarm.com
www.alpacas4u2cfarm.com

Circle C Alpacas
www.circlecalpacas.com

Crafdal Farm Alpacas
crafdalfarm.com

Deer Hollow Alpaca Farm
www.deerhollowfarm.com

Destiny Alpacas
www.destinyalpacas.com

G&W FARM
www.gwfarmanimals.com

Georgian Oaks Suri Alpaca Farm
georgianoaksfarm.com

Jerae's Unity Alpacas
www.jeraesunityalpacas.com

Lasso the Moon Alpaca Farm
www.alpacamoon.com

Mays Hill Alpacas
mayshillalpacas.com

Nutt Farm Suri Alpacas
www.nuttfarm.com

Southern Estate Alpacas
www.southernestatealpacas.com

Thunder River Suri Alpacas
www.thunderriveralpacas.com

Walnut Knoll Farm
www.walnutknollfarm.com

Idaho

Aspen Alpaca Company
www.aspenalpacas.com

Paintbrush Alpacas
www.paintbrushalpaca.com

Rising Star Alpacas
www.risingstaralpacas.com

Sweet Pines Alpacas
sweetpinesalpacas.com

Treasure Valley Alpacas
www.treasurevalleyalpacas.com

Illinois

BrushWalker Alpacas
www.bwalpacas.com

Circle B Alpacas
circlebalpacas.com

Keva Ranch Alpacas
www.kevaranchalpacas.com

Lakada Farm Alpacas
www.lakada.com

Odelia Farms
www.odeliafarms.com

SafeHouse Farm Alpacaswww.safehousefarmalpacas.com

The Midnight Moon Alpaca Ranch
www.themidnightmoonalpacaranch.com

Tiskilwa Farms Alpacas
www.illinoisalpacas.com

Willow Glenn Farm
www.willowglenalpacas.com

Wineaux Alpacas
www.wineauxalpacas.com

Wisdom Of The Fox Alpacas
www.wisdomofthefox.com

Indiana

Catch-a-Cria Alpacas
www.catchacria.com

Heritage Farm Suri Alpacas
ourheritagefarm.com

Lookout Farm Huacaya Alpacas
www.lookoutfarmalpacas.com

Mt. Tabor Alpaca Farm
mttaboralpacas.com

Alpacas of Indiana
www.questalpacas.com

Shootin' Stars Farm: Suri Alpacas
shootinstarsfarm.com

Sundance Suri Farm
sundanceSuri.com

Whispering Willows Alpacas at Fishback Creek Farm
www.whisperingwillowsalpacas.com

WindSwept Farms Alpacas
www.wsfalpacas.com

Iowa

Alpacas @ Triple Tree
www.tripletree.us

C & M Acres
cmacres.com

Irish Meadows Alpaca Farm
www.irishmeadowsalpacas.com

Loess Hills Alpacas
www.lhalpacas.com

North River Alpacas
www.northriveralpacas.com

Southern Iowa Alpacas Home
www.southerniowaalpacas.com

SunRise Suris Alpaca Ranch
www.sunriseSuris.com

Kansas

Alpacas at Rockyfield Farm
www.rockyfieldfarm.com

Alpacas of Moose Creek Ranch
www.alpacasofmoosecreekranch.com

Cedar Hollow Alpaca Farm
www.cedarhollowalpacas.com

Serenity Hill Farm Alpacas
www.serenityhillfarmalpacas.com

Kentucky

Maple Hill Manor
maplehillmanor.com

Alpacas of the Bluegrass, LLC
www.alpacasofthebluegrass.com

Angel Fleece Suris & Huacaya
www.angelfleecealpacas.com

Brightside Bend Alpacas
www.brightsidebend.com

Capture Your Heart Alpacas
www.captureyourheartalpacas.com

Eagle Bend Alpaca Farm
www.eaglebendalpacas.com

Poppy Moon Farm
poppymoonfarm.com

Red Roof Ranch Alpacas
www.redroofranch.net

The Shepherd's Criations Alpaca Farm
www.alpacajoy.com

Wooly Acres Alpacas
www.woolyacresalpacas.com

Maine

Black Woods Farm Alpacas
blackwoodsfarmalpacas.com

Cloud Hollow Farm
www.chf1752.com

Misty Acres Alpaca Farm
www.mistyacresalpaca.com

Mountain Brook Farm
www.mtbrookfarm.com

OutBaca Alpaca
www.outbacaalpaca.com

Sandy River Alpacas, LLC
www.sandyriveralpacas.com

The Upper Farm
www.theupperfarm.com

Maryland

A Paca Fun Farm
www.apacafunfarm.com

Alpacas Carlos-Finca Escudero
www.alpacascarlos.com

Bell House Alpacas
www.bellhousealpacas.com

Chesapeake Alpacas
www.chesapeakealpacas.com

Flame Pool Alpacas
www.flamepoolalpacas.com

Gray Alpaca Company
www.grayalpacacompany.com

Outstanding Dreams Farm
www.outstandingdreamsfarm.com

Painted Sky Alpaca Farm
www.paintedskyalpacafarm.com

Pax River Alpacas
www.paxriveralpacas.com

Shear Elegance Alpacas
shearelegancealpacas.com

Shepherd's Purse Alpacas
www.shepherdspursealpacas.com

Sugarloaf Alpaca Company
www.sugarloafalpacas.com

Massachusetts

Acorn Alpaca Ranch
www.acornalpacaranch.com

Angel Hair Alpacas
angelhairalpacas.webs.com

Bayns Hill Alpaca Farm
www.baynshillalpacafarm.com

Big Red Acres Alpacas
www.bigredacres.com

Craigieburn Farm Alpacas
alpacas4u.com

Golden Touch Farm
www.goldentouchfarm.com

Great Rock Alpacas
www.greatrockalpacas.com

Happy Hearts Alpaca Farm
www.happyheartsalpacas.com

Island Alpaca Company
www.islandalpaca.com

Maple Brook Farm Alpacas
www.maplebrookfarm.com

Red Barn Alpacas
www.redbarnalpacas.com

Silver Oak Farm
www.silveroakalpacas.com

Sippican River Farm
www.sippicanriverfarm.com

Sunny Knoll Farm Alpacas
sunnyknollfarmalpacas.com

Michigan

AKA Alpacas At Bonny Hill Ranch, LLC
www.akaalpacas.com

Alpaca Adventures of Mid Michigan
www.alpacaadventuresofmidmichigan.com

Alpaca Heights
www.alpacaheights.com

Alpaca Lane of Central Michigan
www.alpacalane.com

Amazin' Grazin' Alpaca Ranch, LLC
www.amazingrazinalpacaranch.com

Avalon Farm Alpacas
www.avalonfarmalpacas.com

Benchmark Alpacas at the Tin Roof Ranch
www.benchmarkalpacas.com

Blendon Pines Alpaca
Ranchwww.blendonpinesalpacaranch.com

Bucks Meadow Alpaca Farm
www.bucksmeadow.com

Bucks Meadow Alpaca Farm
www.bucksmeadow.com

Country Lakes Alpacas
www.countrylakesalpacas.com

Crystal Lake Alpaca Farm
www.crystallakealpacafarm.com

Double D Alpaca
www.doubledalpaca.com

Dream Catchers Alpacas
www.dreamcatcheralpacas.com

Eclipse Alpacas
eclipsealpacas.com

Great Lakes Ranch
www.greatlakesranch.com

Loney's Alpaca Junction
www.lajalpaca.com

M&M'z Alpacas
m-mzalpacas.com

Northern Dreams Alpaca
www.woolandhoney.com

Powder Puff Pacas
www.powderpuffpacas.com

Shady Hollow Suri Alpacas
www.shSurialpacas.com

Sunset Dreams Alpaca Farm
www.sunsetdreamsalpacas.com

Triple Diamonds Alpaca Ranch
www.triplediamondsalpaca.com

Two Branch Ranch
twobranchranch.com

Urban Dreams Farm
www.urbandreamsfarm.com

Whitefeather Creek Alpaca Ranch
www.whitefeathercreekalpacas.com

Zodiac Ranch
www.zodiacranch.com

Minnesota

Alpacas by the Brook
www.alpacasbythebrook.com

AlpacaSus Exquisite Alpacas
www.alpacasus.com

Fossum Family Farm Alpacas LLC
www.fossumfamilyfarm.com

Glacial Ridge Alpacas
www.glacialridgealpacas.com

Jen & Ole's Alpacas & Organics LLC
jenandoles.blogspot.com

Lakeland Alpacas
www.lakelandalpacas.com

Pine Cone Ridge Alpacas
pineconeridgealpacas.com

Spirit Song Alpacas
www.spiritsongalpacas.net

Whispering Oaks Alpacas
whisperingoaksalpacas.org

Zonkers Farm
zonkersfarm.com

Mississippi

Kuska Alpacas
www.kuskapaku.com

Missouri

Blue Feather Farm
www.bluefeatherfarm.com

Casa Del Paca
www.casadelpaca.com

Curly Eye
curlyeye.com

Hasselbring's Harmony Ranch
www.hasselbringsharmonyranch.com

Mariposa Farm Alpacas
www.mariposafarmalpacas.com

Mid Missouri Alpacas
www.applesandalpacas.com

R&B Alpaca Ranch
rbalpacaranch.com

River Bluff Alpacas
www.riverbluffalpacas.com

Sunflower Hills Alpacas
www.sunflowerhillsranch.com

Montana

Cloud Dancer Alpacas
www.clouddanceralpacas.com

Flying Alpacas Ranch
www.flyingalpacasranch.com

Nebraska

River Valley Alpacas
www.rivervalleyalpacas.com

New Hampshire

Sleeping Monk Farm
www.sleepingmonkfarm.com

Contoocook Alpaca
contoocookalpaca.com

Foggy Bottom Ranch Alpacas
www.foggybottomranch.com

Foss Mountain Farm
www.fossmtnfarm.com

Indigo Moon Farm
www.indigomoonfarm.net

Inti Alpacas, LLC
www.intialpacas.com

Purgatory Falls Alpaca Farm
purgatoryfallsalpaca.com

Sallie's Fen Alpacas
www.sfalpacas.com

Snowfield Alpacas
www.snowfieldalpacas.com

Someday Farm
www.somedayfarm.com

Stonewall Fields Alpaca farm
www.stonewallfields.com

New Jersey

Alma Park Alpacas
www.almapark.com

Bay Springs Farm Alpacas
www.bayspringsalpacas.com

Dancing Horse Farm
www.dancinghorsealpacas.com

Emelise Alpacas
www.emelisealpacas.com

Hum-Dinger Alpacas
www.humdingeralpacas.com

Jersey Breeders
www.jerseybreeders.com

Makena Farm
makenafarmalpacas.tripod.com

Meadows Edge Alpaca Ranch
www.meadowsedgealpacas.com

Pepper Pot Farm Alpacas
www.pepperpotfarm.com

Swallow Hill Farm
www.swallowhillfarmalpacas.com

Windy Farm Alpacas
www.windyfarmalpacas.com

New Mexico

Akasha Alpacas
www.akashaalpacas.com

Albuquerque Alpacas
www.albuquerquealpacas.com

Blue Mesa Alpacas
www.bluemesaalpacas.com

Phi Beta Paca: Alpacas of Taos tm
phibetapacaalpacasoftaostm.blogspot.com

Wellspring Suri Alpacas
www.wellspringSurialpacas.com

Windrush Alpacas
www.windrushalpacas.com

New York

Alpacas of Maggie's Brook Farm
www.maggiesbrookfarm.com

Ausable Valley Alpacas
www.ausablevalleyalpacas.com

Blooming Field Farm Alpacas
bffalpaca.com

Brooklyn Alpacas
www.brooklynalpacas.com

Cabin View Alpacas
www.cabinviewalpacas.com

Castle Tower Alpacas
www.castletoweralpacas.com

Claddagh Farm Alpacas
www.claddaghfarmalpacas.com

Close-Knit Alpacas
www.closeknitalpacas.com

Copper Star Alpaca Farm
www.copperstaralpacafarm.com

Crooked-Creek-Alpaca-Farm
www.crookedcreekalpacafarm.com

Crosswind Farm Alpacas
crosswindfarmalpacas.com

Edgewood Farm Alpacas
www.edgewoodalpacas.com

Faraway Farm Alpacas
www.farawayfarmalpacas.com

Finger Lakes Alpacas
www.fingerlakesalpacas.com

Gentle Breeze Alpacas
www.gentlebreezealpacas.com

Hellman's Windy Hill Farm Alpacas
www.windyhillfarmalpacas.com

Hemlock Hills Alpaca Farm
hemlockhillsalpaca.com

Ideuma Creek Alpacas
www.icalpacas.com

Imagine Alpacas!
www.imaginealpacas.com

Jay Mountain Alpacas
www.jaymtnalpacas.com

Lazy Acre Alpacas
www.lazyacrealpacas.com

Little Creek Farm Alpacas
www.lcfalpacas.com

Log Cabin Alpacas
www.logcabinalpacas.com

Majestical Meadows Alpacas
www.majesticalmeadows.com

Mountain Meadows Farm
www.mtnmeadowsfarm.com

Nether Walnut Hill Alpacas
www.netherwalnuthill.com

Preston's Alpacas LLC
www.prestonsalpacasllc.com

Quarry Ridge Alpacas
www.quarryridgealpacas.com

Rosehaven Alpacas
www.rosehavenalpacas.com

Salmon River Alpacas
salmonriveralpacas.com

Sanger-La Alpacas
www.sanger-la.com

Shalimar Alpacas
www.shalimaralpacas.com

Sixth Day Farm
www.sixthdayfarm.com

Stoney Elm Alpacas
www.stoneyelmalpaca.com

Sugartown Farms
www.sugartownfarms.com

Tartan Acres LLC
tartanacres.com

Tinbrook
www.tinbrook.com

Zen Alpaca
zenalpaca.com

North Carolina

Alpaca Dreams, LLC
www.alpacadreamsnc.com

Alpacas at Cherry Run
alpacasatcherryrun.com

Augustyn Acres
www.augustynacresalpacas.com

Beech Springs Alpacas
www.beechspringsalpaca.com

Borderline Farms
www.borderlinefarms.com

Celestine Ridge Alpacas
www.celestineridgealpacas.com

Happy Tails Alpacas
www.happytailsalpacas.com

Keepsake Farm Alpacas
keepsakefarmalpacas.com

Landmark Farm Alpacas
www.landmarkfarmalpacas.com

Majestic View Alpacas
www.majesticviewalpacafarm.com

High Meadow Alpaca Farm
www.ncalpacafarmofhighmeadow.com

Southern Cross Alpacas
www.southerncrossalpacasinc.com

The Alpaca Ranch at Cobb Creek Cabins
www.cobbcreekcabins.com

Two Crows Farm
twocrowsalpacas.com

Venezia Dream Farm
www.veneziadream.com

Walkapaca Farm
www.walkapacafarm.com

Ohio

AJ's Alpaca Ranch
ajsalpacaranch.com

Alpaca Bella Suri Farm, LLC
www.alpacabella.com

Alpaca Green
alpacagreen.com

Alpaca Sunrise Farm
www.alpacasunrise.com

Alpacas at Phoenix Hill Farm
www.alpacasatphoenixhill.com

Alpacas of Chappel Creek
www.alpacasofchappelcreek.com

Arborway Alpaca Farm
arborwayalpacafarm.com

CR Alpacas, Inc
www.c-r-alpacas.com

Cameo Rose Alpacas
www.cameorosealpacas.com

Club Suri Alpacas
www.clubSuri.com

Coffee Pot Farm
www.coffeepotfarm.com

Dewey Morning Alpacas
deweymorningalpacas.com

Dynasty Farms
www.dynastyalpacas.com

Eieio Alpacas
www.eieioalpacas.com

Family Jewels Alpacas
www.familyjewelsalpacas.com

Glacial Ridge Farm
glacialridgefarm.com

Harmony Ridge Farms
www.harmonyridgefarms.com

Heatherbrook Farms LLC
www.heatherbrookfarms.com

Hickory Ridge Alpacas
www.hickoryridgealpacas.com

Hidden Hilltop Alpaca Ranch
www.hhalpaca.com

Kaleidoscope Alpacas
www.kaleidoscopealpacas.com

KB Alpacas
www.kbalpacas.com

Lazy G Alpacas
www.lazygalpacas.com

Margery-Ray Alpacas
www.margery-rayalpacas.com

Mystical Acres Alpacas
www.mysticalalpacas.com

One Fine Day Alpacas
onefinedayalpacas.com

Our Little World Alpacas
www.ourlittleworldalpacas.com

Redwoodranchalpacas.com
www.redwoodranchalpacas.com

Renaissance Farms
www.renfarmsohio.com

Ross Ranch, LLC
www.rossalpacaranch.com

Spring Ridge Alpacas
www.springridgealpacas.com

Stewart Heritage Farm
www.stewartheritagefarm.com

Storybook Alpacas
www.storybookalpacas.com

That'll Do Farm
www.thatlldofarm.com

The Alpaca Rosa
www.thealpacarosa.com

The Alpacas of Phantasy Pharm
phantasypharm.com

Topknot Suri Farm
www.topknotSurifarm.com

Victoria Lane Alpacas
www.victorialanealpacas.com

Whistler's Glen Alpacas
whistlersglen-alpacas.com

Wolf Creek Alpacas
wolfcreekalpacas.com

Yes Suri Alpacas
www.yesSurialpacas.com

Oklahoma

Just Right Alpacas
www.justrightalpacas.com

Zena Suri Alpacas
www.zenaSurialpacas.com

Oregon

AdoraBella Alpacas
abalpacas.com

Marquam Hill Ranch
www.mhralpacas.com

Alpacas at Suri Hill
www.Surihill.com

Alpacas at Tucker Creek
alpacasattuckercreek.blogspot.com

Alpacas of Tualatin Valley
www.alpacatv.com

Cascade Shadow
www.cascadeshadowalpacas.com

Chappell's Alpaca Junction
www.calpacajunction.com

Easy Feelin' Alpacas
www.easyfeelinalpacas.com
Eldora Suri Alpacas
www.eldoraSurialpacas.com

Halo Ranch Alpacas
www.haloalpacas.com

Happy Valley Alpaca Ranch
happyvalleyalpacaranch.com

Heritage Alpacas, LLC
www.heritagealpacas.com

Mahart Alpacas
www.mahartalpacas.com

North Plains Alpacas
www.northplainsalpacas.com

Siskiyou Alpacas
www.Suri-futures.com

Skyline Alpaca Farm
www.skylinealpacas.com

Stonyridge Alpacas
www.stonyridgealpacas.com

Vineyard View Alpacas
www.vineyardviewalpacas.com

Wings And A Prayer Alpacas
www.wingsandaprayeralpacas.com
ZZ Alpacas
www.zzalpacas.com

Pennsylvania

Allegheny Alpacas
www.allalpacas.com

Alpaca Angels Farm
www.alpacaangelsfarm.com

Alpaca Palace LLC
www.alpacapalace.com

Alpaca Ventures
www.alpacaventures.com

Alpacaholic Acres
www.alpacaholic-acres.com

Alpacas of Gettysburg
www.alpacasofgettysburg.com

Alpacas of Menges Mills
www.alpacasofmengesmills.com

Alpacas of the Alleghenies
www.gentlealpacas.com

Alpaca Valley Farms
www.alpacavalleyfarms.com

Artisan Alpacas LLC
www.artisanalpacas.com

Asgard Acres Alpaca Farm, LLC
www.asgardacresalpacas.com

Aurora Alpacas & Llamas
www.auroraalpacasllamas.com

Backstage Alpaca Shop
www.backstagealpaca.com

Bent Pine Alpaca Farm
www.bentpinealpacas.com

Cider Press Alpacas
www.ciderpressalpacas.com

Eastland Alpacas
www.eastlandalpacas.com

Falls Edge Farm & Fiber Mill
www.fallsedge.com

Flying Pony Alpacas, LLC
www.flyingponyalpacas.com

Four Points Alpacas, LLC
www.fourpointsalpacas.com

Harley Hill Farm LLC
www.harleyhillfarm.com

Hart-So-Big Alpaca Farm
www.hartsobigalpacafarm.com

Heather's Acre Alpaca Farm
www.heathersacrealpacafarm.com

Heaven's Hill Alpacas
www.heavenshillalpacas.com

Hideaways Heavenly Acres
www.hideawayalpacas.com

Highland Alpaca
www.highlandalpaca.com

Hillside Alpacas
www.hillsidealpacas.com

Laurel Highlands Alpacas
www.laurelhighlandsalpacas.com

Morning SKY Farm
www.morningskyfarm.com

Nobility Alpacas
www.nobilityalpacas.com

Over Home Alpacas: Home
www.overhomealpacas.com

Patchwork Farm Alpacas
www.patchworkfarmalpacas.com

Penncroft Alpacas
www.penncroftalpacas.com

Perkiomen Creek Ranch
www.perkiomencreekranch.com

Quarry Critters Alpaca Ranch
www.quarrycrittersalpacas.com

Rainbow Mountain Alpacas
www.rainbowmountainalpacas.com

Shasta Springs Alpacas
www.shastaspringsalpacas.com

Silkie's Farm
www.silkiesfarm.com

Silvercloud Farm Alpacas
www.silvercloudfarm.com

Snake Creek Alpaca Farm
www.snakecreekalpacafarm.com

Spring Grove Alpaca Ranch
www.spring-grove-alpaca-ranch.com

Stone Meadow Alpacas
www.stonemeadowalpacas.com

Sunnybrook Alpacas
www.sunnybrookalpacas.com

Sunrose Alpacas
www.sunrosealpacas.com

Sweet Valley Suris
www.sweetvalleySuris.com

Terrace Mountain Alpacas, LLC
www.terracemountainalpacas.com

The Wood Farm
www.thewoodfarmalpacas.com

Tuscarora Alpaca Ranch
www.tuscaroraalpacaranch.com

WestPark Alpacas
westparkalpacas.com

Woodland Hills Alpacas
www.woodlandhillsalpacas.com

Tennessee

AlpacaCreekFarm
www.alpacacreekfarm.com

Appalachian Alpacas
www.appalachianalpacas.com

High Meadow Alpacas
www.highmeadowalpacas.net

Humming B Alpacas
www.hummingbalpacas.com

Long Hollow Suri Alpacas
www.longhollowalpacas.com

Mistletoe Farm Alpacas
www.mistletoefarmalpacas.com

Tanasi Trace Alpacas
www.tanasitracealpacas.com

Tennessee Valley Alpacas
tnvalleyalpacas.com

Texas

Ace in the Hole Ranch
www.aceintheholeranch.com

Birch Knoll Alpacas
birchknollalpacas.com

Cherry Bud Farm
www.cherrybudfarmalpacas.com/

Cibolo Creek Alpaca Ranch
www.cibolocreekalpacas.com

Fox Island Alpacas
www.foxislandalpacas.com

McPaca Ranch
www.mcpacaranch.com

R & R Alpacas
www.randralpacas.com

Rancho Paloma
www.ranchopalomaalpacas.com

Reiling Ranch Alpacas
www.reilingranchalpacas.com

Royal Oaks Alpacas
www.royaloaksalpacas.com

Sylvester Ranch Alpacas
sylvesterranchalpacas.com

Trinity Ridge Alpacas
www.trinityridgealpacas.com

Utah

Sierra Bonita Alpacas
www.alpacautah.com

Virginia

7 Springs Alpaca Farm
www.7springsalpaca.com

Alpaca Pastures of Virginia, Inc
www.alpacapasturesva.com

Alpacas of Foster Knoll
www.alpacasoffosterknoll.com

Morning Mist Alpacas
morningmistalpacas.com

Alpacas of Tranquility
www.alpacasoftranquility.com

Alpacas Plus of Virginia
www.alpacasplusofva.org

Box Elder Ranch, LLC
www.boxelderranch.com

Broad Creek Alpacas
broadcreekalpacas.tripod.com

Buttonwoodalpacas.com
www.buttonwoodalpacas.com

Cameron Mountain Alpacas
www.cameronmountain.com

CrimpHaven Alpacas
www.crimphaven.info

Diamond Triple C Ranch
www.diamondtriplecranch.com

Double "O" Good Alpacas
www.doubleogood.com

Double JJ Alpacas
www.doublejjalpacas.com

Enchanted Hill
www.enchantedhillalpacas.net

Heartline Alpaca Farm
www.heartlinealpacafarm.com

Heronwood Farm Alpacas
www.heronwoodfarm.com

Little Wing Alpaca Farm
www.alpacalove.com

Mayhem Farm
www.mayhemfarm.com

MoLi Ranch, Inc.
www.themoliranch-alpacas.us

Ore Hill Farm Alpacas
www.orehillfarmalpacas.com

Otter Peaks Alpacas
otterpeaksalpacas.com

Paca Criations
www.pacacriations.com

Peaceful Heart Alpacas
peacefulheartalpacas.com

Poplar Hill Alpacas
www.poplarhill.com

Ranch Acres Alpacas
ranchacresalpacas.com

Rivanna River Farm, LLC
www.rivannariveralpacas.com

River Mist Farm Alpacas
www.rivermistfarm.net

Scenic View Alpaca
scenicviewalpaca.com

Suri Downs Farm
www.Suridownsfarm.com

Take Me Home Alpacas
www.takemehomealpacas.com

Wildwood Alpacas
www.wildwoodalpacas.com

Vermont

Cas-Cad-Nac Farm
www.cas-cad-nacfarm.com

Log Cabin Farm Alpacas
www.logcabinfarm.com

Maple View Farm Alpacas
www.mapleviewfarmalpacas.com

Mystic Meadow Alpacas
www.mysticmeadowalpacas.com

North of the Andes Alpaca Farm
www.noaalpacas.com

Parris Hill Farm Alpacas & The AlpacArt
www.parrishillfarm.com

Vermont Alpaca Company
www.vermontalpacaco.com

Wayfarer Farm
www.wayfarer-farm.com

Wild Apple Alpacas
www.wildapplealpacas.com

Wildwood Acres Alpacas
www.wildwoodacresalpacas.com

Washington

3D's Alpacas
www.3dsalpacas.com

Aleutian Eagle Alpacas
www.aleutianeaglealpacas.com

Alpacas at Paradise Pointe
www.alpacasatparadisepointe.com

Alpacas of Wintercreek
www.alpacasofwintercreek.com

Apple Country Alpacas
www.applecountryalpacas.com

Black Hills Alpacas
www.alpacahappy.com

Fern Ridge Alpacas
www.fernridgealpacas.com

Genesis Alpacas
www.genesisalpacas.com

Kettle Ridge Alpacas
kettleridgealpacas.com

Morning Mist Alpacas
morningmistalpacas.com

Sawdust Hill Alpacas
sawdusthillalpacas.com

Shadow Ridge Alpacas
www.shadowridgealpacas.com

Southfork Suri Ranch
www.southforkSuriranch.com

Spring Canyon Alpacas
springcanyonalpacas.com

Starshire Ranch
www.starshireranch.com

Thunder Mountain Alpacas
www.thundermountainalpacas.com

Yakima River Alpacas
yakimariveralpacas.com

West Virginia

Briar Run Alpacas
www.briarrunalpacas.com

West Virginia Alpacas
www.kismetacres.com

Morgan's Fortunato Farm
www.fortunatoalpacas.com

Orchard Hill Alpacas
www.alpacasofwv.com

Shepherds Creek Alpacas
www.shepherdscreekalpacas.com

Wisconsin

Amazing Grace Alpaca Farm of Wisconsin
www.amazinggracealpacaswi.com

Double D Alpaca Ranch
www.doubledalpacaranch.com

Enchanted Meadows Alpacas
www.enchantedmeadows.com

GalPaca Farm
www.galpacafarm.com

Hidden Pond Farm, LLC
www.hiddenpondfarmalpacas.com

Huacaya Hills Alpacas
www.Huacayahillsalpacas.com

Kinney Valley Alpacas
kinneyvalleyalpacas.com

Lodi Alpacas
www.lodialpacas.com

Lotis Alpacas
www.lotisalpacas.com

Oak LawnAlpacas Farm
www.oaklawnalpacas.com

Offbeat Acres Alpaca Farm
www.offbeatacres.com

Peaceful Pastures Alpacas
www.peacefulpasturesalpacas.com

Sauk Creek Alpacas
www.saukcreekalpacas.com

Sugar Creek Alpacas, LLC
www.sugarcreek-alpacas.com

Timber Ridge Farm
www.timberridgefarm.com

Token Creek Alpacas
www.tokencreekalpacas.com

Wyoming

Amber Sky Alpacas
www.amberskyealpacas.com

Lone Tree Alpacas LLC.
www.lonetreealpacas.com

UK Alpaca Farms

Apollo Alpacas
Call: 01295 713188
Near Banbury in the Heart of England.
Email: apolloalpacas@btinternet.com
Website: www.apolloalpacas.co.uk

Beacon Alpacas
Husthwaite, North Yorkshire.
Tel: 01347 868879 or 07716 917315
Email: info@beaconalpacas.co.uk
Website: www.beaconalpacas.co.uk

Chalfield Alpacas
Wiltshire close to the city of Bath.
Tel: 07786 228647
Email: janet@chalfieldalpacas.co.uk
Website: www.chalfieldalpacas.co.uk

Herts Alpacas
Hertfordshire
Tel: 01763 271301 or Mobile 07802 433155
Email beckwith904@aol.com
Website: www.hertsalpacas.co.uk

Ivywell Alpacas
South Gloucestershire
Telephone: 07772222179
Email: alpacas@ivywell.com
Website: www.ivywell.com

Kensmyth Stud Alpacas
Near Cirencester in the Cotswolds.
Tel: Helen on: 07799700587 or 01285 862020
Email: Helen@kensmyth.com
Web: www.kensmyth.com

Lightfoot Alpacas
Hawkhurst in the Weald of Kent.
Tel: 07802 263589
Email: graham@alpacabreeder.co.uk
Website: www.alpacabreeder.co.uk

Lyme Alpacas
Lyme Regis
Tel: 07887 511774
Email: sue.thomas@lymealpacas.co.uk
Website: www.lymealpacas.co.uk

Pennybridge Alpacas
North Hampshire, close to the Surrey and Berkshire borders.
Tel: 01256 764824 or Mob: 07801 132757
Email: joy@pennybridgealpacas.co.uk
Website: www.pennybridgealpacas.co.uk

Scotfield Alpacas
Warfield, Berkshire
Tel: Sue Hipkin 07770 455533
or Lisa Batup 07770 455534
Website: www.scotfieldalpacas.co.uk

Snowshill Alpacas
North Cotswold close to Snowshill.
Tel: 01386853841/ 07711044106
Email: roger.mount@snowshillalpacas.com
Website: www.snowshillalpacas.com

Toft Alpacas
Located on the shores of Draycote Water.
Tel: 01788 810626 or 07970 626245
Email: shirley@toft-alpacas.co.uk
Website: www.toftalpacastud.com

West Wight Alpacas
Five minutes from the Yarmouth ferry port, which is only
half an hour's crossing from Lymington in Hampshire.
Tel: 01983 760900
Website: www.westwightalpacas.co.uk

Canada

7 Heaven Alpacas
www.7heavenalpacas.ca

Alpagas du Massif
www.alpagasdumassif.com

Alpage du Nord
www.alpagedunord.com

Arriba Linea Alpacas
www.arribalinea.com

Beniuks Alpacas
www.beniuksalpacas.com

Camelot Haven Alpaca Ranch
www.camelothaven.com

Highgate Alpacas
www.highgatealpacas.com

Hilltop Oasis Alpacas
www.hilltopoasisalpacas.ca

Hubbert Alpaca Farms
www.hubbertfarms.ca

KJ Alpacas
www.kjalpacas.com

Living Sky Alpacas
www.lsalpacas.ca

Ring Ranch Alpacas
www.ringalpacas.com

Tocino Alpacas
www.tocinoalpacas.com

Twistlane Alpacas
www.twistlanealpacas.com

Afterword

From their lofty prominence in Incan culture to their "re-discovery" during the 19th century Industrial Revolution, alpacas have long lived in association with man, providing humans with one of the finest and most durable of all natural fibers.

Easily kept and highly adaptable to climate and region, alpacas are suitable pets and livestock for a wide range of ownership models. A pair can be easily kept by a fiber artisan who wants to harvest the fleece annually for small projects, or multiple animals will graze peacefully in the pasture of a larger operation requiring little in the way of special fencing and shelter.

Male "herdsires" with impeccable pedigrees make excellent investments, earning their keep in stud fees. Often these animals are bought by groups of investors, since as many as 80% of all alpacas are gelded in an effort to constantly improve the gene pool. The relative scarcity of herdsires dramatically inflates the prices for these prized animals.

Genetic registries are maintained by alpaca groups around the world, and there is a concerted effort by dedicated breeders to recreate the superior level of fleece once cultivated by the Incans.

Although alpaca is unquestionably one of the great luxury fibers of the world, the fossil record tells us that this marvelous material achieved an even greater level of development before 90% of the existing population was killed by the Spanish Conquistadores.

I hope that you will come away from this text having found a model for alpaca ownership that fits your goals and lifestyle. As with any kind of animal, especially livestock, you will enter the field a frightened novice, and in short order gain a working and comfortable knowledge of alpacas and their husbandry.

One thing I perhaps haven't said in the text is that alpacas, at least in my experience, are genuinely likeable. When treated kindly and well, they are agreeable to handling and individual animals are often quite affectionate. At shows and exhibitions I've watched young children work with alpacas with complete ease and comfort.

I do believe that it's extremely important for "newbies" in the industry to attend alpaca shows, to make friends in the industry, and to cultivate mentoring relationships with them. To really learn the ins and outs of life with alpacas, you need someone to talk with, to bounce ideas around with, and to consult with when you're just plain confused.

Beyond this relationship, I also suggest you make certain that you have access to a knowledgeable veterinarian before you ever purchase your first alpaca. With these key players in place, you'll have a much easier time in your first year of alpaca ownership.

Regardless of the road you take, I can assure you that your association with alpacas will enrich your life. Legend holds that these placid creatures are on loans from Pachmana, the Earth Mother, and they are, without question a position gracious gift to we mere humans.

Relevant Websites

Please note that the following websites were extant in mid-2014 at the time of this writing. Due to the ever changing nature of the Internet, no guarantee can be made that these addresses will be valid in the future.

Alpaca Culture: Alpaca News and Information from Around the World
alpacaculture.com

Alpaca World Magazine
www.alpacaworldmagazine.com

The Alpaca Owners Association, Inc.
www.alpacaregistry.com

Alpaca Seller: Matching Buyers and Sellers Across the World
www.alpacaseller.com

British Alpaca Futurity
www.britishalpacafuturity.com

Alpaca Research Foundation
www.alpacaresearchfoundation.org

Camelid Identification System (CIS)
www.cisdna.org

International Camelid Institute
www.icinfo.org

Alpaca Owners Association, Inc. (AOA)

alpacainfo.com

Australian Alpaca Association (AAA) & (IAR) Registry

www.alpaca.asn.au

Australasian Alpaca Breeders Association Inc. (AABA)

www.aaba.com.au

Lama and Alpaka Register (Austria)

www.lamas.at

Canadian Llama and Alpaca Association and Registry (CLAA)

www.claacanada.com

Alpaca Canada

www.alpacainfo.ca

European Suri Association

www.europeanSuriassociation.com

Alpaca Breeders of Finland

www.alpakkakasvattajat.fi

Alpagas et Lamas de France & Registry (l'AFLA)

www.alpagas-lamas-france.org

Alpaka Zucht Verband Deutschland e.V. & Registry (AZVD)

www.alpaka.info

Societa Italiana Alpaca (SIA)

www.sialpaca.it

Alpaca Association of New Zealand (AANZ) & (IAR) Registry

www.alpaca.org.nz

The Norwegian Alpaca Association

www.alpakkaforeningen.no

International Alpaca Association (IAA) (Peru)

www.aia.org.pe

South Africa Alpaca Breeding Society

www.alpacasociety.co.za

Llama and Alpaca Registries Europe (LAREU) (Switzerland)

www.lareu.org

Alpaka Verein Schweiz (VLAS), Alpaca Association Switzerland

www.vlas.ch

The British Alpaca Society and British Alpaca Registry (BAS)

www.bas-uk.com

Suri Network

www.Surinetwork.org

Glossary

A

aggregate breeding value - An animal's breeding value for a combination of desirable traits. Also called "net merit."

agouti - The gene for "wild" or "natural" color. In alpacas, this is the "vicuna" color.

altiplano - The high plateau area around Lake Titicaca in southern Peru and northwestern Bolivia this is the natural range of the alpaca.

apron - The fiber under the neck and around the chest of an alpaca. This is coarse fiber that should not be mixed with the finer fiber from other parts of the body.

artificial insemination - The process by which collected semen from stud males is inserted into female alpacas in either frozen or fresh form.

B

bale - Sacks, squares, or packages of compressed alpaca fiber that vary in size.

beater - The spiked or toothed metal roller of an opening or cleaning machine used in the processing of alpaca fiber into yarn.

blanket - The high quality fleece from the shoulder of an alpaca extending to the area along the back and down each side where it meets with, but does not include, the belly fiber. The blanket also does not include the fiber from the neck, chest, legs, or britch.

bloodline - A term that refers to the pedigree of an alpaca through a traceable ancestry.

blowout factor - The degree to which the diameter of an individual animal's fiber increase or thickens over time.

breeding objective - An established goal for a breeding program according to a definition or list of the qualities that makes for an optimum animal and herd.

C

Camelidae - Animals classed as "camelids," including camels, alpacas, llamas, guanacos, and vicunas.

carding - The final cleaning process, done either by hand or by machine, before alpaca fiber is ready to be spun.

comb - Aligning alpaca fibers by either hand or machine combing which prepares the material to be spun into worsted yarn.

conformation - A term that describes an alpaca's body in terms of shape, appropriate alignment, and balance.

cria - The correct term for an alpaca off less than one year of age. Pronounced "cre-a."

crimp - An even undulation along the length of alpaca fiber that has the appearance of corrugation. The more crimps per inch / 2.54 cm, the finer the fiber.

crinkle - The term that describes the even, corrugated wave present in a single Huacaya fleece fiber.

culling - The process of determining which animals within a herd are not suitable for breeding.

curl - Spiraling ringlets of lustrous Suri fiber that gives the coat what is called a "drenched" appearance.

cuticle - The outer cell layer of alpaca fiber. Each cuticle cell is hard and flat. They do not fit evenly together, but stand out from the shaft with a serrated edge.

D

dam - The female in a mated pair of alpacas.

density - The given number of fibers present in a specific area of an alpaca's body.

F

felt - A technique for the creation of a sheet of matted material held together by the entanglement of the fibers, not by weaving. The joining is accomplished through a combination of heat, moisture, and pressure.

fertility - Refers to a male's ability to impregnate a female or her ability to conceive and produce offspring.

fineness - A measure of the diameter of individual alpaca fibers expressed in microns as an average of a representative sample. Generally varies from 20-36 microns, with lower numbers indicating finer or "better" fibers.

follicle - The structure on the skin from which an individual fiber or hair grows.

fleece - The single coat of one animal sheared intact from the animal once a year.

fleece weight - The weight of the whole fleece taken from an animal and measured at the same time each year.

G

gelding - A male alpaca that has been castrated and thus rendered incapable of siring offspring.

grading - Sorting fibers by staple length, strength, evenness, and fineness for the purpose of classification.

greasy - A term used in the industry to refer to alpaca wool that has not yet been washed.

guard hair - A coarse, secondary coat fiber found in alpacas. Sometimes referred to as "kemp."

H

hair fiber - Fiber obtained from animals other than sheep, but included in the term "wool." Typically this material comes from goats and members of the camel family including alpacas.

hembra - The correct term for a female alpaca.

handle - Term to describe the feel of alpaca fiber when touched. Synonymous with the term "softness."

hank - A defined length for a textile that varies by type. For materials like alpaca fiber classed as wool, a hank is 560 yards / 512 meters. For cotton and silk, the length is 840 yards / 768 meters.

herdsire - A male alpaca with desirable genetic characteristics for breeding into a herd.

histogram - A group of data points organized in a graphical representation that quantifies the most desirable characteristics of alpaca fiber.

humming - The sound an alpaca makes when it is happy or curious, but one that can also indicate discomfort or concern depending on the prevailing circumstances.

Huacaya - Alpaca breed displaying a well-crimped fleece that grows perpendicular to the skin.

Huarizo - A derogatory term that describes a cross between an alpaca and a llama. The result is an animal of poor breed type with weak and coarse, modulated fiber.

hybrid vigor - A marked improvement in the health of hybrid animals over that of purebreds. Some traits that tend to be most noticeable include survivability and fertility.

I

impurity - When used in reference to a fleece or textile, an impurity refers to any extraneous or undesirable material present.

K

kemp - A term referring to guard hairs or medullated fibers.

keratin - The protein that is the main structural component of woolen fibers.

kush - When an alpaca sits down with its legs bent under its body to rest, the position is called "kush."

L

loft - A term that is roughly synonymous with "fluffiness," and refers specifically to the ability of a fiber to "spring"

---

back to its normal position once it has been squeezed or compressed.

loom - A mechanical or hand-operated device used to make woven cloth.

luster - A kind of glisten or sheen that is highly desirable in alpaca fiber.

M

macho - A term used in reference to a male alpaca.

maiden - The term for a female alpaca of breeding age who has not yet been bred.

maternal trait - Qualities like fertility, milk production, and the ability to mother offspring that are especially desirable when breeding female alpacas.

mean fiber diameter - The measurement of the average diameter (thickness) of a group of alpaca fibers.

medulla - The hollow core typical of guard hairs in the chest and underbelly of an alpaca fleece.

melanin - Pigment that is responsible for both skin and coat color.

micron - A measurement equal to 1/25,000 of an inch, or 1/1000th of a millimeter. Microns are used to express fiber diameter, which is a reference to the fineness of the fiber. The smaller the micron number, the finer the fiber.

multicolor - An alpaca with more than one color present in the fleece.

N

nep - Refers to a small knot of short, dead, and immature fibers that have become tangled.

noil - Refers to the short fibers that are removed when fiber is combed to make yarn.

O

open female - An alpaca female not bred currently.

orgling - A term used to describe the mating sound of a male alpaca that helps to induce ovulation in the female.

P

pedigree - The registered or recorded line of descent that forms the genealogy of an alpaca's ancestry.

piebald - An alpaca marked with white and black patches. May also be called a pinto.

production sequence - The sequential procedure to produce fiber, which includes: shearing, sorting, opening, cleaning, carding, drawing, combing, roving, twisting, and spinning.

prime fleece - The best or highest quality fleece one alpaca will ever produce. Often this is the first coat, which is called tui.

pronking - A term that describes how an alpaca "skips." The animals hops with all four feet hitting the ground simultaneously. The behavior often occurs at dusk, and indicates that the alpaca is happy. It is seen most often in cria, but older alpacas do it as well.

puna - A zone of barren tundra high in the Andes Mountains where alpacas live naturally.

pureblood - An animal bred from a recognized type with no mixed ancestry.

R

resilience - Also referred to as memory, resilience is the ability of a fiber or yarn to spring back into it original state once the force that caused the deformation (crushing or wrinkling) has been removed.

registry - A center for the storage and retrieval of information about alpacas including pedigree data, blood typing, registry numbers, and other vital information.

roan - Alpaca coat that contains a uniform mix of colored fibers.

roving - Cleaned and carded fleece that is then drawn out to make a twisted roll. The fiber can then be used as it is for felting or spun into yarn.

S

scouring - The process of cleaning raw fiber to remove impurities. These may include dirt, sweat, and grease. Scouring is accomplished with soaps or chemicals.

second-cut - An instance of careless shearing that necessitates the need to re-shear areas that were not taken down to the skin the first time. Greatly diminishes the worth of the fleece.

shearing - The annual harvesting of alpaca fiber, typically done in mid-spring. The process keeps the animal cool in the summer and allows the coat ample time to come back in before winter.

shear - The removal of the alpaca's fleece in one large piece (blanket) done with hand or electric shears.

sire - The father in a mated pair of alpacas. Sometimes referred to as the "herdsire."

skewbald - An alpaca marked with white and brown patches. Sometimes called a pinto.

skirt - The removal by hand of contaminants from a shorn alpaca fleece.

spinning - The twisting of fiber into yarn. The process may be accomplished with a drop spindle, a spinning wheel, or commercial machinery.

staple length - The measurable length of a lock of shorn alpaca fiber.

staple - A group or cluster of individual alpaca fibers. A large group of staples form a fleece.

stud - A male alpaca used for the purpose of breeding.

Suri - A type of alpaca with long "dreadlocks" or "pencil" curls of non-crimped fiber. The locks lie close to the animal's body and twist vertically downward.

T

tags - The waste wool on the shearing floor that is discarded, generally broken or covered in dung.

tender wool - A staple of fiber that can only be carded but not combed due to weak areas attributable to illness, exposure to the elements, or poor diet.

tensile strength - The degree of pulling force that can be applied to a fiber before it stretches and then breaks.

tippy wool - A fleece with badly weathered ends that contains significant amounts of grease, dirt, and debris.

tui - A term referring to an alpaca aged 6-18 months or to the fleece it produces, which will likely be the best and highest quality the animal will ever grow.

twist - In a length of yarn, the number of turns about the axis per unit. Usually expressed as "turns per inch" or tpi.

V

vicuña - The small, South American camelid that is considered the direct ancestor of the alpaca. Vicuñas have cinnamon and white coats that are the finest natural fiber found in the world.

W

weanling - An alpaca less than one year old that has been weaned and is no longer drawing nutrition from nursing its mother.

weaving - The process by which yarn is interlaced at right angles according to a pattern to make cloth.

wool cap - The wool on the head and between the ears of an alpaca, which is considered to be desirable in terms of aesthetics. Also called a topknot or bonnet.

woolen - Yarn that is bulky and thick, having been made from fibers 1-3 inches / 2.54 – 7.62 cm long that have been carded, with no further processing.

worsted - Yarn made from alpaca fibers 3 inches / 7.62 cm or more long that has been carded, combed, and drawn.

Y

yearling - An alpaca that is 1-2 years of age.

yolk - A colorless impurity made of grease and sweat found in fiber.

Index

Index

Made in the USA
Middletown, DE
11 September 2017